Domestic Trends

in the United States, China, and Iran

Implications for U.S. Navy Strategic Planning

John Gordon IV, Robert W. Button, Karla J. Cunningham,
Toy I. Reid, Irv Blickstein, Peter A. Wilson, Andreas Goldthau

T0146310

Prepared for the United States Navy

 NATIONAL DEFENSE RESEARCH INSTITUTE

The research described in this report was prepared for the U.S. Navy's Office of the Chief of Naval Operations, Assessment Division (N81). The research was conducted in the RAND National Defense Research Institute, a federally funded research and development center sponsored by the Office of the Secretary of Defense, the Joint Staff, the Unified Combatant Commands, the Department of the Navy, the Marine Corps, the defense agencies, and the defense Intelligence Community under Contract W74V8H-06-C-0002.

Library of Congress Cataloging-in-Publication Data

How domestic trends in the U.S., China, and Iran could influence U.S. Navy strategic
    planning / John Gordon IV ... [et al.].
        p. cm.
    Includes bibliographical references.
    ISBN 978-0-8330-4562-1 (pbk. : alk. paper)
    1. United States. Navy—Planning. 2. Strategic planning—United States.
3. Social prediction—United States. 4. Economic forecasting—United States.
5. Social prediction—China. 6. Economic forecasting—China. 7. Social
prediction—Iran. 8. Economic forecasting—Iran. I. Gordon, John, 1956–

VA58.4.H69 2008
359'.030973—dc22

                                                                    2008044846

The RAND Corporation is a nonprofit research organization providing objective analysis and effective solutions that address the challenges facing the public and private sectors around the world. RAND's publications do not necessarily reflect the opinions of its research clients and sponsors.

RAND® is a registered trademark.

Cover Photo Credits: AP Images/Greg Baker; AP/Vahid Salemi
Cover Design by Carol Earnest

Published 2008 by the RAND Corporation
1776 Main Street, P.O. Box 2138, Santa Monica, CA 90407-2138
1200 South Hayes Street, Arlington, VA 22202-5050
4570 Fifth Avenue, Suite 600, Pittsburgh, PA 15213-2665
RAND URL: http://www.rand.org/
To order RAND documents or to obtain additional information, contact
Distribution Services: Telephone: (310) 451-7002;
Fax: (310) 451-6915; Email: order@rand.org

Preface

This book examines the future of the United States, the People's Republic of China, and Iran. Specifically, it reviews how important domestic trends—such as changes in the economy, demographics, and the environment—may influence the priorities of these three nations in the future. This research is the second in a series of strategic studies the RAND Corporation conducted for the U.S. Navy's Office of the Chief of Naval Operations, Assessment Division (N81). The initial research was conducted in the summer and fall of 2006. Entitled "Evolving Strategic Trends, Implications for the U.S. Navy," that first study was intended for a select Navy audience. It identified likely major global strategic trends in the next decade and how they might influence Navy planning. Although the present book focuses primarily on domestic trends, it also explores how events in the so-called near abroad of each nation may influence how the three principal nations view their strategic situation. Based on the insights gained during their research, the authors provide conclusions and recommendations that will be of interest not only to the study's sponsor, the U.S. Navy, but also to a wider range of policymakers and academics in the United States and elsewhere.

This research was sponsored by the U.S. Navy's Office of the Chief of Naval Operations, Assessment Division (N81), and conducted within the International Security and Defense Policy Center (ISDP) of the RAND National Defense Research Institute, a federally funded research and development center sponsored by the Office of the Secretary of Defense, the Joint Staff, the Unified Combatant Commands, the Department of the Navy, the Marine Corps, the defense agencies, and the defense Intelligence Community.

For more information on RAND's International Security and Defense Policy Center, contact the Director, James Dobbins. He can be reached by email at James_Dobbins@rand.org; by phone at 703-413-1100, extension 5134; or by mail at the RAND Corporation, 1200 South Hayes Street, Arlington, Virginia 22202-5050. More information about RAND is available at www.rand.org.

Contents

Figures

Tables

Summary

This monograph is the second in a series of strategic studies conducted by the RAND Corporation for the U.S. Navy's Office of the Chief of Naval Operations, Assessment Division (N81). The initial research was conducted in the summer and fall of 2006. Entitled "Evolving Strategic Trends, Implications for the U.S. Navy," that first study was intended for a select Navy audience. It identified likely major global strategic trends in the next decade and how they might influence Navy planning. As a result of that study, N81 asked RAND to conduct a follow-on effort that focused primarily on the domestic trends of the United States, China, and Iran. The Navy wanted insights on how these important trends could influence U.S. security decisions in general and the Navy's allocation of resources in particular.

Study Approach

Whereas the first strategic-trends study was primarily concerned with security-related issues around the world that could influence U.S. military planning, this monograph discusses internal, nonmilitary trends in the United States, the People's Republic of China (PRC), and Iran. The Navy is interested in these countries' likely key "resource drivers" from now through roughly 2020–2025.[1] Accordingly, we examined important domestic trends in each country—in demographics,

---

[1] The period covered by data related to future economic and demographic projections for the United States, China, and Iran varied. In some cases, projections through 2020 were

economics, energy consumption, the environment, and education—to gain an understanding of each nation's likely "big issues." Depending on how much of a challenge those issues become in the United States, China, and Iran, the Navy may have to divert considerable resources to address emerging problems.

Although our primary focus is on domestic trends in each nation, this monograph also examines each nation's so-called near abroad. We conducted this research to determine how much of a challenge each of the three nations will experience in their own immediate "neighborhoods." In the case of the United States, this neighborhood includes the Caribbean, Central America, Mexico, and northern South America. We divided China's near abroad into three general regions: the east-northeast (where most of China's near-term security challenges lie), the south and southeast, and the west. In Iran's near abroad, the Middle East, we examine how the current turbulent situation might influence Iran's strategic planning and resource-allocation decisions.

N81 also asked that we examine the near abroads of Russia and Japan. This was considered important due to each country's relationship with the three primary countries. For example, China closely watches how Japan's security policy is evolving, and also has a very important relationship with Russia. Our assessments of Russia and Japan do not include the more-detailed research on domestic trends we conducted for the three primary nations.

The United States

The United States will remain the richest nation in the world during the period covered in this monograph. Today, the U.S. economy is roughly $13 trillion. That total is projected to rise to roughly $30 trillion by 2020. Thus, most of the U.S. population will continue to enjoy rising standards of living.

---

available; in other cases, projections extended out through 2025. Therefore, this study's "out years," the far-term planning horizon, are 2020–2025.

The main challenge that the United States will experience from now until 2025 (and beyond) will arise from the increasing numbers of elderly Americans. Today, there are approximately three active workers for each recipient of Social Security. That ratio will drop to two workers per retiree by about 2015. Elderly people require considerably more health care and other social services compared to younger portions of the population. Social Security expenditures will increase during this period, but the main increase will occur in Medicare and Medicaid, the federal government's health programs for those over the age of 65. Although the U.S. economy is projected to expand annually from now until 2025, social spending for America's growing numbers of elderly citizens will consume an ever-increasing portion of the federal budget and the overall economy. This increase will constrain other spending programs, including national defense. After 2015, given this projected need for an increase in social spending for the elderly, and absent a clearly perceived threat comparable to that posed by the Soviet Union during the Cold War, it is likely that senior U.S. policymakers will be less willing—and possibly less able—to devote as much of the nation's wealth to defense as they did during the Cold War or even today's global war on terrorism.

Despite the coming challenge of increased social spending as "baby boomers" enter their retirement years, the United States is a rich nation that will be able to provide for increasing numbers of elderly citizens (if it chooses to do so). In that regard, "the United States became rich before it became old." We will see that a different situation prevails in China.

The United States has also been fortunate that it has not had to devote considerable military resources to its near abroad since the Spanish-American War of 1898. Although there have been occasional periods of tension (such as the Cuban Missile Crisis of the early 1960s), the United States has generally been able to keep its military focused in other, more-distant regions. Whether that favorable situation will continue depends in large part on (1) whether there is a sustained increase in the popularity of anti-American, leftist regimes in Latin American (such as Hugo Chavez's Venezuela) and (2) events in post-Castro Cuba, where a chaotic situation could result in some form of protracted U.S. military intervention.

China

The PRC has enjoyed 30 years of explosive economic growth. This economic growth has enabled China to undertake a significant military modernization program for the past several years. In the near to medium term (i.e., through roughly 2020), the growth of China's economy will continue to enable the Chinese to expand the nation's military capabilities. Projections of China's gross domestic product (GDP) in 2020 vary considerably, ranging from $12 trillion to $15 trillion. We argue that mounting internal pressures will limit China's ability to expand militarily into the out years.

China has entered into a Faustian bargain, however. In exchange for wealth and military power, China has sacrificed its energy (coal) future, embarked on a program of unsustainable economic growth, sacrificed the well-being of its elderly, and allowed extensive environmental damage to occur. These problems are intertwined. The depletion of coal reserves, for example, will encourage the Chinese to use lignite coal, a fuel source that produces less energy and more pollution than higher grades of coal. Increased use of lignite would increase pollution and thus lead to increased acid rain and environmental degradation.

Until recently, China has been self-sufficient in coal. It now uses coal for about 70 percent of its electrical power. Coal consumption in China has roughly kept pace with China's economic growth, increasing by 80 percent since 2000 and 2,000 percent since China's last publicly released coal survey in 1992. Several recent credible studies suggest that China's coal production will peak sometime between 2015 and 2025, with coal production levels between 2030 and 2040 falling below current levels unless significant new reserves are found. Even if new reserves are located, China could face significant challenges in extraction, coal quality, and infrastructure. In addition to significantly increasing its oil and natural gas imports, China is transitioning from a net exporter of coal to a net importer.

Note, however, that some of the data underlying these figures are uncertain. For example, coal-reserve estimates could be low, and coal is still being discovered in China.[2] Furthermore, growing coal shortages could encourage both exploration for additional sources of coal and greater coal production efficiency. Some coal geologists argue that China has a 2,500-year history of organized coal production, making further major coal finds unlikely. Coal engineers point to problems in increasing the efficiency of mines that have not achieved acceptable safety levels. Clearly, exponentially increased consumption rates have correspondingly advanced the date of China's coal peak. What appears certain is that as the years go by, China will have to import a larger amount of coal to meet its requirements. This greater dependence on foreign coal could affect international coal prices. (However, because coal is more globally plentiful than oil, increased Chinese coal imports are not likely to affect coal prices.) Greater dependence will probably result in increasing Chinese ties with the nations from which it imports coal (e.g., Australia).

Until recently, China was self-sufficient in oil; in fact, China was actually an oil exporter until 1993. Oil production in China has not yet peaked, but China's demand for oil under a burgeoning economy has outstripped production. China's urbanization, together with the increased affordability of automobiles, has accelerated and will continue to accelerate China's oil demand. As a newcomer to the international oil market, China first turned to lesser producers (such as Oman) and sought exclusive drilling rights as a means to secure oil supplies. More recently, China has turned to Africa and Iran for oil. Perhaps deliberately, China has, through its choice of oil suppliers, avoided competition or confrontation with the United States in the international oil market. Current trajectories of oil consumption suggest, however, that this pattern will break as China turns to the Greater Middle East for oil. Competition for access to affordable energy sources could become a source of friction in Sino-U.S. relations.

---

[2] Historically, however, China's coal-reserve estimation technology has significantly overestimated coal reserves elsewhere in the world.

China's economy has consistently experienced the world's highest growth rate since the early 1980s, posting an average growth of 10 percent, according to official figures. Although this growth is remarkable, it is not unprecedented. Other Asian economies, such as those of Taiwan, South Korea, and Japan, demonstrated similarly strong growth at the peak of their development trajectories; most of these economies then experienced subsequent slowdowns to more-sustainable growth levels. Meanwhile, the sheer scale of China's internal needs is daunting, and the gap between China and the most-developed nations is clear. For example, China's 2005 GDP per capita was roughly $1,700, a level comparable to that of the United States in 1850.

China's economic progress is complicated by the unique challenges that the country faces as the world's most populous nation. It is also complicated by the fact that China is simultaneously developing and transitioning its economy from a closed, communist command system to a market-driven, open system. Factors that once contributed to China's past growth—including greater openness to trade, improved technology, and a large, youthful labor pool—will contribute less to China's economic growth in the future. Continued reform will require the Chinese Communist Party (CCP) to cede more control to the private sector, thereby weakening the CCP's role in the economy. The CCP has refused to cede control in some major areas, such as coal production, thereby creating economic distortions throughout the economy. Knowing that deeper economic reforms will entail political risk, it is not clear how much farther the CCP will be willing to go in economic reforms. Furthermore, the following existing structural impediments and habitual problems are expected to weaken China's future economic growth:

- The dysfunctional banking system supports approximately 150,000 unprofitable state-owned enterprises that suck resources from enterprises that may be more efficient and profitable.
- China is overly dependent on exports, overinvestment, and undervalued currency for growth.
- Inadequate educational systems produce unqualified graduates.

- China's poor protection of intellectual property rights discourages innovation.

China's once-youthful labor force is aging, and the declining number of active workers will become an economic liability. More generally, China's aging population will present major challenges in the medium to far term (i.e., from 2015 to roughly 2025). China's baby boomers were born in the 1960s, and therefore its working-age population will begin to shrink after 2015. In 2034, in some regions of China, dependents will outnumber active workers.

In transitioning from a command-driven economy to a free-market economy, China has failed to establish a well-funded social security system. Over 300 million Chinese agricultural workers have no social security, and urban workers' social security accounts are underfunded by more than $1 trillion. Today, there are approximately seven active workers for every retiree in China. By 2020, that ratio will drop to only two workers per retiree. Meanwhile, and in spite of the fact that the Chinese economy will continue to grow, China's per capita GDP increases will still leave the nation rather poor by the standards of the industrial world. Compared to the United States, then, it can be said that "China will get old before it gets rich."

By 2025, more than 300 million Chinese citizens will be age 66 or older. This will place a great strain on the nation's social services, which today are considerably behind those of the United States, Japan, and most of Europe. Chinese workers have limited access to health insurance, and preventative health care for children is lagging. More importantly, China's medical system is wholly unprepared to serve its aging population. The strain that an increased number of elderly Chinese will place on China's resources after 2020 will almost certainly affect on the nation's ability to continue a major military buildup.

China has serious and growing environmental problems. Water is scarce and polluted. Two-thirds of China's cities now experience significant water shortages. Over half of all urban waste water is untreated, and over three-quarters of river water in urban areas is unsuitable for drinking or fishing. Many of China's rivers are polluted to levels incomprehensible to most Westerners; these waters are deemed too polluted for any beneficial use, even industrial use. Economic growth and the degradation of China's rivers have increased demands for groundwater. China's deserts are spreading: More than a quarter of China's land is now desert. Groundwater in some regions of China is expected to be exhausted within 15 years, leaving residents without usable water. Economic growth will be retarded by a growing "bow wave" of environmental damage (which is estimated to be slowing China's current GDP growth by about 5 percent).

The rapid industrialization of many of the country's urban areas has entailed high levels of air pollution. These levels are much higher than would be tolerated in the United States or most of Europe. As observed above, China has dramatically increased the use of coal for electrical power to meet its rapidly growing energy needs; it plans to open scores of new coal-fired power plants in the coming years. China's coal is relatively dirty and the country has not invested adequately in coal-cleaning facilities. China's approximately 400,000 coal-fired industrial facilities nationwide produce millions of tons of ash and sulfur dioxide, a source of acid rain.

By 2020, China will enter a "perfect storm" of economic, environmental, and social problems largely of its own making. In the next 10–15 years, while trying to grow and transform its economy, China will confront the intertwined problems of premature depletion of its energy resources, faltering economic growth, inadequate provisions for its aging population, and the need to remediate an extensively damaged environment. China's ability to modernize and expand its military at the same time will be constrained by these domestic challenges.

The PRC's current relations with nations in its near abroad vary considerably. Chinese relations with countries to the south (e.g., Vietnam and India) are better than at any point in the past three decades. Mutual economic interest has contributed greatly to this improved situation. Of significant importance is China's increasing westward strategic orientation into Central Asia and the Middle East. In the near to medium term (i.e., from today through approximately 2020), China will increase its involvement in the former Soviet republics of Central Asia and further expand its growing presence in the Middle East, including Iran. China's rapidly expanding energy needs guarantee that this trend will continue for the foreseeable future. Areas east and northeast of China pose the greatest threat to China in the near to medium term. How the situation with Taiwan evolves in the next decade will have great bearing on Sino-Japanese and Sino-U.S. relations. In addition to the still-uncertain situation with Taiwan, there is potential for a crisis in northeast Asia, where the Stalinist regime of North Korea has acquired nuclear weapons. The Japanese are watching developments in Taiwan and the Korean Peninsula with considerable interest, and a significant change in Japan's security policy is possible. The PRC is likely to take seriously any Japanese actions that may threaten Chinese interests.

Iran

Iran is rich in oil and natural gas. It also has one of the best primary and secondary education systems in the Muslim world. It therefore has the means to develop a healthy economy. However, major structural problems in its economy and political system are causing the country to perform far below its theoretical potential.

Iran's government has been largely ineffective in harnessing the country's energy potential. The country's high unemployment and underemployment rates have resulted in rising numbers of jobless youths. The lack of employment opportunities has contributed to the country's ongoing "brain drain," and tens of thousands of its middle and educated classes depart the nation each year in search of better opportunities elsewhere. Over the next 20 years, Iran's per capita income will grow only modestly.

Iran's inability to adequately exploit its economic potential means that Iran will lack the ability to create a military force that can dominate the entire Middle East. However, as long as current ideology continues to dominate Iran's government, Iran will remain generally opposed to U.S. interests in the region. More likely than any direct confrontation with the United States and European nations is the possibility that Iran will attempt to undermine the U.S. position in the region through more-clandestine means (such as support to terrorist groups and proxies, including Hezbollah). Additionally, Iran will seek out a major ally in the form of Russia, China, or both. If either nation strongly backs Iran, Iran will be more likely to become emboldened in its willingness to challenge U.S. interests in the region. Iran's pursuit of nuclear weapons, and its possible success in that regard, is another issue that could destabilize the entire region.

If Iran's economy continues to underperform, the role of the Iranian military could increase. This could lead to a situation roughly similar to the role of the armed forces in Pakistan or Turkey, which dominate many aspects of government. In the case of Iran, an important issue would be the relative strength of the Revolutionary Guard and the conventional branches of the military.

Japan

Japan will remain the richest nation in Asia through the out years.[3] However, the country's population will age considerably. Unlike the United States, but like China, Japan admits few immigrants. With low immigration levels and a very low birthrate, Japan's population levels are bound to decline in the coming decades. This demographic trend will force changes in social spending in Japan, as will be the case in the United States and China.

Of primary concern for this study were Japan's relations with its near abroad. In that regard, Japan has three major strategic options:

- strengthening its ties with the United States—this is Japan's present course of action
- reaching an accommodation with the PRC—a strategy that acknowledges that U.S. power in Asia is waning, while China's is rising
- adopting a more independent course of action in terms of foreign and security policy—this option may ultimately entail acquiring a nuclear-deterrent capability.

---

[3] As previously noted, RAND examined Japan's near-abroad issues only.

Japan's choice will be of very great interest to the United States and China.

Russia

Following a near-disastrous decade after the collapse of the Soviet Union, Russian strength is starting to rise again.[4] Russia is currently benefiting from the rise in oil and natural gas prices, which has facilitated renewed modernization efforts by the Russian military. Meanwhile, Russia is taking an increasingly assertive approach to its own near abroad.

Of vital interest to the United States is how Russian relations with Iran and China evolve in the coming decade. Iran clearly wants support from Russia, China, or both. Meanwhile, the evolving Sino-Russian relationship could develop in a way that opposes U.S. interests. Russia is also becoming increasingly assertive with former Soviet republics, especially those in Central Asia. This might lead to tension with China, since the PRC is leaning westward due to its growing energy needs.

Implications for the U.S. Navy

Although the United States will remain the world's richest nation for the next 20–30 years, the steady aging of its population will create major shifts in federal spending in the future. In the absence of a clearly perceived threat comparable to the Cold War–era Soviet Union, spending will gradually shift away from defense and toward the increased social services needed to support an older population.

Meanwhile, U.S. dependence on foreign energy sources will continue to grow. Thus, the Navy will be required to continuously maintain powerful forces in the Middle East and the Western Pacific.

---

[4] As previously noted, RAND examined Russia's near-abroad issues only.

Depending on the nature of the threats in those regions, the kind of naval forces required could vary considerably.

The Navy faces uncertainty about the degree to which it will have to prepare for (1) a "high-end" future conflict against a powerful, well-armed opponent (e.g., China) versus (2) the so-called Long War against rogue nations and terrorist organizations (e.g., Iran and various radical nonstate groups). In the latter case, the Navy will have to invest more heavily in amphibious, coastal patrol, and sea-basing capabilities that are conducive to supporting Army, Marine Corps, and coalition forces that are engaged in irregular warfare ashore. This uncertainty about the future produces the classic "blue-water" versus "green/brown-water" investment conundrum that the Navy has faced in previous years. If the situation in the U.S. near abroad deteriorates, the Navy may also find itself having to devote more resources in that direction.

The Navy will therefore have to balance its investment decisions around the following realities:

- As the years go by, increasingly powerful budgetary and political forces in the United States will tend to downplay military spending.
- Absent a clearly perceived foreign threat, willingness to spend considerable sums on defense will decrease over time, and resources will instead be directed toward social spending.
- The United States will need to acquire the appropriate numbers and types of naval forces to prepare for both of the two possible scenarios (the high-end conflict and the Long War).
- The Navy will have to continue providing forward presence in a variety of missions in regions critical to the United States. This obligation could include dealing with the implications of shifting alliance and coalition arrangements in critical regions.

Many of the Navy's decisions will depend on the evolution of U.S. relations with China and Iran and the direction of the Long War during the next ten years. Given the long lead times associated with Navy programs and the decades-long life span of Navy platforms once constructed, the effects of the Navy's important near-term decisions will endure well beyond the out years.

Acknowledgments

We wish to express our thanks to CAPT John Yurchak, CDR Brian Clark, and LCDR Harrison Schram of N81, the sponsors of this study. Our frequent, frank, exchanges with them were of great value. N81 was very open to the insights RAND developed during the course of the research, an attitude that was much appreciated by the study team. Various colleagues at RAND were also very helpful. In particular, Dr. Keith Crane provided us with a considerable amount of data about Iran, and Mr. Charles Wolf and Mr. Daniel Byman provided helpful reviews of the draft manuscript.

Abbreviations

BP	British Petroleum
Btpa	billions of tonnes per annum
Btu	British thermal unit
C4ISR	command, control, communications, computers, intelligence, surveillance, and reconnaissance
CASS	Chinese Academy of Social Sciences
CBO	Congressional Budget Office
CCP	Chinese Communist Party
CO_2	carbon dioxide
CRS	Congressional Research Service
DoD	U.S. Department of Defense
DoE	U.S. Department of Energy
DPRK	Democratic People's Republic of Korea
E&P	exploration and production
EIA	Energy Information Administration
ELN	*Ejército de Liberación Nacional* [National Liberation Army]
EU	European Union

EWG	Energy Watch Group
FARC	*Fuerzas Armadas Revolucionarias de Colombia* [Revolutionary Armed Forces of Colombia]
FDI	foreign direct investment
FSU	Former Soviet Union
FY	fiscal year
FYDP	Future Years Defense Plan
GAO	Government Accountability Office
GDP	gross domestic product
GME	Greater Middle East
IAEA	International Atomic Energy Agency
IEA	International Energy Agency
IPR	intellectual property rights
IRGC	Islamic Revolutionary Guard Corps
ISDP	International Security and Defense Policy Center
IT	information technology
mbd	millions of barrels per day
MOH	PRC Ministry of Health
Mtoe	million tonnes of oil equivalent
N81	Office of the Chief of Naval Operations, Assessment Division
NGO	nongovernmental organization
NOAA	National Oceanic and Atmospheric Administration
NPC	National Petroleum Council

NPL	nonperforming loan
OECD	Organisation for Economic Co-operation and Development
OMB	Office of Management and Budget
PCA	EU-Russia Partnership and Cooperation Agreement
PLA	People's Liberation Army
PPP	purchasing-power parity
PRC	People's Republic of China
PRD	*Partido de la Revolución Democrática* [Party of the Democratic Revolution]
PRI	*Partido Revolucionario Institucional* [Institutional Revolutionary Party]
R&D	research and development
RMB	renminbi
ROK	Republic of Korea
SARS	severe acute respiratory syndrome
SEPA	PRC State Environmental Protection Agency
SOE	state-owned enterprise
TPEC	total primary energy consumption
TVE	township and village enterprise
UN	United Nations
WEO	World Energy Outlook

CHAPTER ONE

Introduction and Objectives

Introduction

This research is the second in a series of strategic studies RAND conducted for the U.S. Navy's Office of the Chief of Naval Operations, Assessment Division (N81). The initial research was conducted in the summer and fall of 2006. Entitled "Evolving Strategic Trends, Implications for the U.S. Navy," that first study was intended for a select Navy audience. It identified likely major global strategic trends in the next decade and how they might influence Navy planning. As a result of that study, N81 asked RAND to conduct a follow-on effort that focused primarily on the domestic trends of the United States, China, and Iran. Ultimately, the Navy wanted insights on how these important trends could influence U.S. security decisions in general and the Navy's allocation of resources in particular.

Study Approach

Whereas the first strategic-trends study was primarily concerned with security-related issues around the world that could influence U.S. military planning, this monograph discusses internal, nonmilitary trends in the United States, the People's Republic of China (PRC), and Iran. The Navy is interested in these countries' likely "resource drivers" from now through roughly 2020–2025.[1] Accordingly, we examined important domestic trends in each country—in demographics, economics, energy consumption, the environment, and education—to gain an understanding of each nation's likely "big issues." Depending on how much of a challenge those issues become in the United States, China, and Iran, the Navy may have to divert considerable resources to address emerging problems. For example, our research clearly demonstrates that the U.S. and Chinese populations are "graying." The elderly require considerably more health care and other social services resources than the younger portions of a population. As the percentage of older citizens rises in these countries, therefore, there will be a need to reallocate national resources in that direction. This shift could have a significant effect on resources that would otherwise be available for national defense.

Although our primary focus is on domestic trends in each nation, this monograph also examines each nation's so-called near abroad. We conducted this research to determine how much of a challenge each of the three nations will experience in their own immediate "neighborhoods." In the case of the United States, this neighborhood includes the Caribbean, Central America, Mexico, and northern South America. This is, of course, a vitally important region for the United States, but it is an area where, since the Spanish-American War, the United States has not had to devote considerable military resources. Should relations with its southern neighbors worsen considerably, the United States might be forced to increase its security presence in the region, possibly at the expense of other commitments elsewhere in the world.

We divided China's near abroad into three general regions: the east-northeast (where most of China's near-term security challenges lie), the south and southeast, and the west. In Iran's near abroad, the Middle East, we examine how the current turbulent situation might influence Iran's strategic planning and resource-allocation decisions.

---

[1] The period covered by data related to future economic and demographic projections for the United States, China, and Iran varied. In some cases, projections through 2020 were available; in other cases, projections extended out through 2025. Therefore, this study's "out years," the far-term planning horizon, are 2020–2025.

N81 also asked that we examine the near abroads of Russia and Japan. This was considered important due to each country's relationship with this monograph's three primary countries. For example, China closely watches how Japan's security policy is evolving, and also has a very important relationship with Russia. Our assessments of Russia and Japan do not include the more-detailed research on domestic trends conducted for the three primary nations.

The Navy asked that we develop near-, medium-, and far-term insights about how key trends will affect the United States, China, and Iran. We assume a near-term period of approximately five years, which roughly corresponds to the U.S. Department of Defense's (DoD's) five-year plan. We further define a five-year medium term (approximately 2015–2020) and a ten-year far term (2020–2025). Past studies demonstrate that the farther into the future one tries to extrapolate trends, the less accurate the forecasts tend to be. In most trend areas (e.g., the economy, energy, the environment) we therefore present each trend's general direction and its likely implications for each of the three primary nations. In the case of China, for example, current pollution problems in many Chinese urban areas are already severe. However, the PRC intends to significantly increase the number of coal-fired power plants in the next decade. Therefore, we project that China's environmental problems will continue to worsen (at least in the near to medium term) unless the PRC implements a significant policy shift, such as regulations that impose much stricter standards on its coal-fired power plants. We also describe strong evidence that recent increases in China's use of coal will transform China from a net exporter of coal to a net importer. Accordingly, China will increasingly depend on outside sources of oil, natural gas, and coal to meet its energy needs.

There is one major area where the trend projections are firm: demographics. Setting aside the possibility of a major pandemic or other cataclysmic event that could radically alter a nation's demographic trajectory, it is possible to project population trends 30 to 40 years into the future with considerable accuracy. In that regard, and given this monograph's focus on domestic trends in the primary countries, demographics are one of the most important areas that we examined. As mentioned above, projections clearly show that the U.S. and Chinese populations are graying, and that fewer young workers will be available to support the considerable social services needs of ever-greater numbers of elderly citizens. The numbers are sufficiently staggering to make it likely that each nation will have to significantly reallocate resources at the national level by 2020. Meanwhile, Iran will still be experiencing a "youth bulge" through 2025 and for some years beyond.

Organization of This Monograph

Examining each of the three primary countries in turn, we first review domestic trends, highlighting insights about the economy, demographics, energy needs, and other factors. We then present a short chapter on each nation's near abroad. Following these reviews and two chapters that examine the neighborhoods of Japan and Russia, we provide an overall assessment of the trends and show how they could influence each primary nation's allocation of resources at the national level. We conclude with implications for the U.S. Navy.

CHAPTER TWO

Strategic Trends in the United States

Summary

In discussions about defense affordability, the defense budget of the United States is frequently measured as a percentage of gross domestic product (GDP).[1] This chapter argues that strategic trends in the United States will reduce the relevance of thinking about the DoD budget as a percentage of GDP. The DoD budget as a percentage of discretionary spending will, however, become more relevant. The inevitable graying of the U.S. population will increase nondiscretionary federal spending at rates that exceed economic growth. With greater pressure being exerted on discretionary spending as a whole, the DoD budget will be squeezed regardless of the DoD budget as a percentage of GDP. Structural factors will reinforce this trend. That said, defense spending will continue to rise as the U.S. economy expands, but it will likely fall as a percentage of GDP due to the need to devote an increasing amount of resources to other areas. Plausible or even likely "wild-card" factors may further pressure defense spending. In this chapter, we estimate when additional pressure will be applied to the DoD budget and identify the portion of the DoD budget most likely to be affected by limits on discretionary spending. To permit country comparisons at the end of this monograph, we supply additional material, including projections of per capita GDP and energy supply and demand.

---

[1] The Heritage Foundation, for example, has argued that U.S. defense spending as a percentage of GDP is too low. See The Heritage Foundation, *Federal Revenue and Spending: A Book of Charts*, undated.

Introduction

Our research for this chapter began with a broad review of those trends within the United States that are likely to affect the DoD budget. We found many interesting overall trends. For example, in the United States, men in their 30s earn less today compared to their fathers' generation, and family income growth has slowed.[2] The most pervasive trends, those in which projections are most confident, relate to the graying of the American population.

U.S. Demographic Trends

Birthrates in the United States spiked between 1946 and 1964 (Figure 2.1), producing the so-called baby boom. Birthrates then sagged below the replacement rate of approximately 2.1 children per female, but have since recovered to meet the replacement rate. Therefore, any significant growth in the population of the United States will occur through immigration.

Life expectancy at birth in the United States took an upturn early in the 20th century. The life expectancy of males at birth increased progressively from 48 years in 1900 to 75 years in 2003. In the same period, the life expectancy of women at birth increased progressively from 51 years to 80 years. The life expectancy of both sexes at age 65 took an upturn in the middle of the century. Life expectancy of males at age 65 has increased from 12 years to 17 years in approximately the past 50 years. The life expectancy of women at age 65 has increased from 12 years to 20 years. Improved access to health care, advances in medicine, healthier lifestyles, and better health before age 65 explain decreased death rates among older Americans. In combination with rising life expectancies, the aging of the baby boomer generation can be viewed as a tipping point in American demographics. Figure 2.2 shows age groups as a percentage of the overall population. Age groups start at the bottom of the figure with the very young (less than five years old)

---

[2]   Isabel Sawhill and John E. Morton, *Economic Mobility: Is the American Dream Alive and Well?* Washington, D.C.: The PEW Charitable Funds, 2007.

Figure 2.1
Historical and Projected U.S. Birthrates

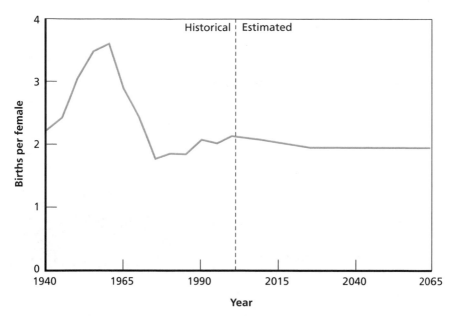

SOURCE: Social Security Administration, *The 2002 Annual Report of the Board of Trustees of the Federal Old-Age and Survivors Insurance and Disability Insurance Trust Funds*, March 26, 2002.
RAND *MG729-2.1*

and increase at the top of the figure to the very old (80 or more years old). A striking feature of this display is the increase in the population of Americans age 65 or older. In 1990, about 8 percent of the American population was 65 or older. By 2050, that age group is expected to represent about 21 percent of the population. The two vertical bars in Figure 2.2 bookend the period from 2011 to 2029, when baby boomers will reach age 65. The percentage of Americans over 65 will increase most rapidly in this period, and then stabilize at about 20 percent. Simultaneously, the percentage of Americans who are 80 or more years old will more than double (from 3.8 percent to 8 percent of the population). There will be no return to the status quo after these changes. Instead, after this period, there will be a new demographic status quo.

Today, 49 million U.S. citizens receive Social Security benefits from the federal Old-Age, Survivors, and Disability Insurance pro-

Figure 2.2
Long-Term U.S. Demographic Trends

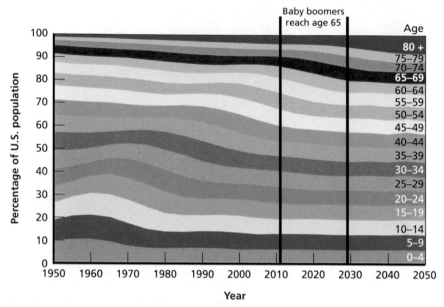

SOURCE: U.S. Department of Commerce, Census Data, 2007.
RAND MG729-2.2

gram. About 70 percent of those beneficiaries are retired workers and their dependents. The remaining beneficiaries are survivors of deceased workers or disabled workers and their dependents. Trends in population aging are therefore less useful for predicting Social Security costs than they are for predicting health costs. We turn now to beneficiaries of and contributors to Social Security.

There are now about 3.3 workers for each Social Security beneficiary. By 2030, this ratio is expected to decline to 2.2 workers for each beneficiary (Figure 2.3).[3] This is the root source of problems, discussed later, that will affect Social Security.

---

[3] Social Security Administration, *The 2007 Annual Report of the Board of Trustees of the Federal Old-Age and Survivors Insurance and Disability Insurance Trust Funds*, May 1, 2007, p. 10.

Figure 2.3
Number of Workers per Social Security Beneficiary

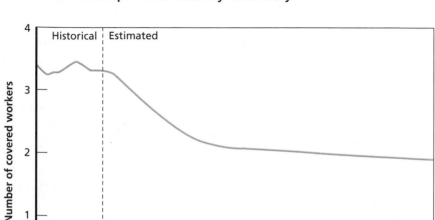

SOURCE: Social Security Administration, *The 2007 Annual Report of the Board of Trustees of the Federal Old-Age and Survivors Insurance and Disability Insurance Trust Funds*, May 1, 2007, p. 10.
RAND MG729-2.3

Gross Domestic Product

GDP forecasts for the U.S. economy universally project uninterrupted growth for the indefinite future. GDP forecasts out to 2030 from the Bureau of Economic Analysis are shown in Figure 2.4. GDP forecasts out to 2017 from the Congressional Budget Office (CBO) are consistent with the Bureau of Economic Analysis GDP forecast and are referenced later in this chapter.

The GDP growth projected in Figure 2.4 combines with slow population growth to yield projections about per capita GDP (Figure 2.5).

As noted previously, continuity is a key assumption in GDP (and federal spending) forecasts. This chapter describes two discontinuities that could adversely affect GDP, federal spending, or both. It also

Figure 2.4
GDP Growth Projection

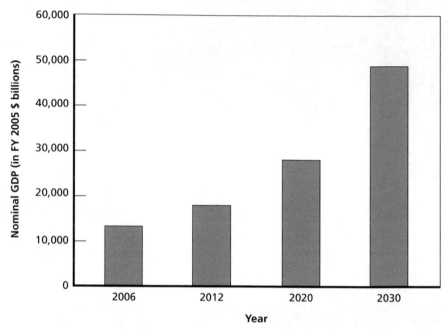

SOURCE: Bureau of Economic Analysis, "National Economic Accounts," last updated
August 28, 2008.
RAND MG729-2.4

describes the economy's dependence on foreign energy supplies and the
possible consequences of an interruption in the United States' imported
oil supply.

The Federal Budget

This monograph's financial and budgetary projections are predicated
on the continuation of current laws and policies. The projections are
presented to illustrate expected sources of pressure on the DoD budget
and to identify when these pressures might exert their influence.

Figure 2.5
Projected per Capita U.S. GDP

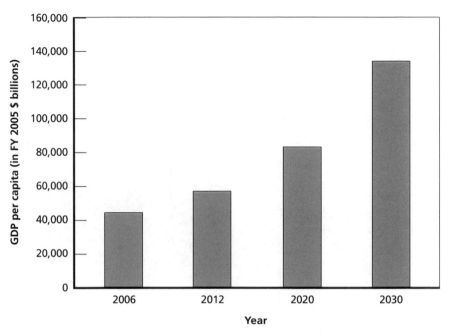

SOURCE: Economist Intelligence Unit, "Market Indicators and Forecasts," Web page, 2007a.
RAND *MG729-2.5*

Federal Revenues

Federal revenues as a percentage of GDP have varied little since 1951, ranging from 16.1 percent to 20.8 percent. Federal revenue is likely to remain in this range.

Federal Spending

As shown in Figure 2.6, overall federal spending has accounted for nearly 20 percent of GDP for the past 40 years. The CBO believes that increasing the federal budget to 25 percent of GDP could jeopardize fiscal stability. Mandatory programs, primarily Social Security, Medicare, and Medicaid, have grown significantly as a share of the federal budget (and as a percentage of GDP). Nondefense discretionary spending has stubbornly resisted change as a share of the federal budget.

Figure 2.6
Historical and Near-Term Federal Spending as a Percentage of GDP

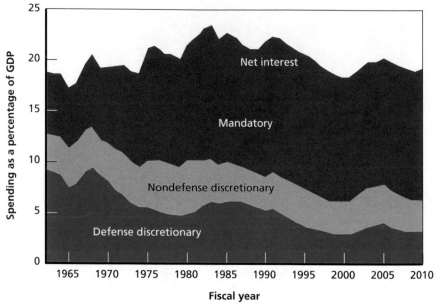

SOURCE: S. Daggett, *Defense Budget: Long-Term Challenges for FY2006 and Beyond*, Congressional Research Service, Order Code RL32877, April 20, 2005, p. 4.
RAND *MG729-2.6*

Social Security

The combined assets of the Social Security trust funds now total $2 trillion, equal to 3.45 times the current annual expenditure rate. The funds' board of trustees projects that by 2016, combined assets will total $4.2 trillion, equal to 4.07 times the rate of expected expenditures in that year.[4] Social Security costs will increase more rapidly than tax revenues between 2010 and 2030 due primarily to the retirement of baby boomers. This pattern will be extended beyond 2030 by increases in life expectancy (note that fertility rates will remain stable at near the replacement rate). Annual Social Security expenditures are expected to start exceeding tax revenues in 2017. Trust-fund exhaustion is projected in 2041. Several measures could provide stability: an increase of

---

[4]   Social Security Administration, 2007, p. 2.

1.95 percentage points to the combined payroll tax rate; a 13-percent reduction in benefits; general revenue transfers of $4.7 trillion in present value; or some combination of these approaches. Solvency beyond 75 years would require significantly greater changes.

With cash-flow deficits for the Social Security program expected to begin in 2017, funds will then begin coming from the U.S. Treasury's General Fund. Pressure on the federal budget from Social Security is then expected to begin in 2017.

Medicare and Medicaid

National health expenditures have outstripped GDP growth for decades. In 1960, these expenditures were 5.1 percent of GDP; in 2001, they were 14.1 percent of GDP. Both publicly and privately funded health expenditures have outgrown GDP. Several structural factors for rising health expenditures have been identified. For example, technological advances in tests and treatments generally raise costs by increasing demand for services. Widely available public and private health insurance policies give consumers little incentive to restrict their consumption of services.

In this context, Medicare's costs have, for decades, increased even more rapidly than national health expenditures (and GDP). Factoring out demographic trends, Medicare cost per enrollee increased 3 percent faster than per capita GDP between 1970 and 2003.

Since 1975 (the earliest year for which data are readily available), Medicaid's costs have also grown more rapidly than national health expenditures. In this period, cost per beneficiary has increased more than tenfold. In part, this increase reflects optional services added under state plans. As with national health care expenditures, technology has increased Medicaid costs. Prescription drugs have been another source of higher costs. Finally, Medicaid-sharing arrangements between states and the federal government encourage states to pursue federal reimbursements, thereby increasing federal spending for Medicaid.

We believe that the problem of Medicare and Medicaid cost growth is as intractable as the problems associated with Social Security. As noted earlier, the costs of medical services are outstripping inflation. Expensive new technologies and drugs, and increased demand for them, increase health care costs. Additional structural factors are at play. For example, the leveraging effects of insurance policies (with their fixed deductibles and co-payments) contribute to the problem. Cost-shifting practices, which allocate the costs of care delivered to one population through above-cost revenue collected from another patient population, reduce the need for efficiency in the delivery of medicine. Additional government-mandated benefits and other legislative changes increase costs. The utilization of medical services is increased by the development and promotion of improved diagnostic services. All of these factors compound the effects of the expected increase in the utilization of medical services due to aging.

Many of the demographic and structural factors expected to drive up civilian health care costs will apply to military health costs as well (Figure 2.7).

Figure 2.7
U.S. Military Medical System Spending by Category

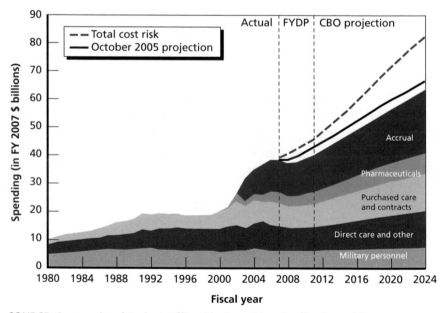

SOURCE: Congressional Budget Office, *The Long-Term Implications of Current Defense Plans: Detailed Update for Fiscal Year 2007*, April 2007, p. 10.
RAND *MG729-2.7*

Interest on the National Debt

The budget scenarios used by the CBO do not incorporate the economic effects of a rising national debt. Nonetheless, the CBO has analyzed the likely effects of mounting federal debt. It concludes that growth of the federal debt over time is not necessarily a problem. As long as GDP growth outstrips the growth of the federal debt (i.e., as long as the ratio of GDP to federal debt does not decrease), there will be no problem. However, GDP growth may not keep pace with growing federal debt levels. At some point, the economy may not support the government's need to service the debt. Sustained excessive debt would affect the economy by siphoning funds from the nation's savings pool and by reducing investments. Investments in business structures, equipment, research and development (R&D), and worker training and education would be reduced, thereby slowing the growth of worker productivity and real wages and ultimately affecting GDP. Rising debt could eventually lead to a sustained contraction of the economy. Eventually, these developments could stop foreign investments in U.S. securities, devalue the dollar, stoke inflation, create recession, collapse stock markets, and seriously weaken the economies of U.S. trading partners.[5]

Federal Spending

Under current laws and policies, Social Security costs will start increasing quickly in approximately 2011 as baby boomers begin to reach the age of 65. Medicare and Medicaid expenses will continue to ramp up as more Americans enroll in those programs and as the medical needs of the aging population increase (see Figure 2.8). Increases in Social Security costs can be controlled to some degree, but the problem of a declining ratio of workers to Social Security recipients cannot be eliminated. Worse, as shown in Figure 2.8, intractable Medicare and Medicaid cost growth will dominate Social Security as a source of cost growth. Under current laws and policies, total federal spending will reach a record percentage of GDP in approximately 2035, and this record level may hurt economic growth. As shown in Figure 2.8, deficits and interest on debt

---

[5] Congressional Budget Office, *The Long-Range Budget Outlook*, December 2003.

Figure 2.8
Federal Outlays by Category, 1950–2075

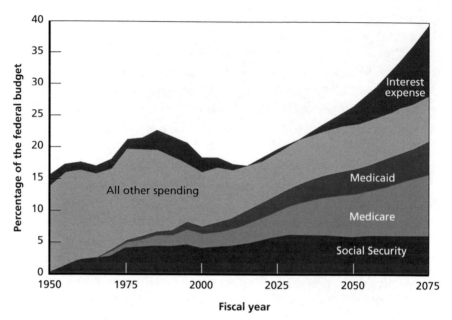

SOURCE: Congressional Budget Office, *A 125-Year Picture of the Federal Budget's Share of the Economy*, Long-Range Fiscal Policy Brief, revised July 2002a.
RAND MG729-2.8

will mushroom and, before 2075, overtake Medicare and Medicaid as federal budget items.

Discretionary Spending

Discretionary spending is that portion of the federal budget that is negotiated between the President and Congress each year as part of the budget process. Defense discretionary spending includes DoD, homeland security, and the war on terror. Of $2.655 trillion in outlays in fiscal year (FY) 2006, discretionary spending totaled $1.032 trillion. The Office of Management and Budget (OMB) has determined that in FY 2006, $532 billion was allocated for defense spending and $500 billion was allocated for nondefense spending.[6]

Nondefense Discretionary Spending

Nondefense discretionary spending funds Social Security, Medicare, Medicaid, education, income security, health research and regulation, highways and mass transit, justice administration, international affairs, natural resources and the environment, and other categories. Accounting for about 18 percent of federal spending in FY 2006, an amount equal to 4 percent of GDP, nondefense discretionary spending represents less than $5 billion, and can provide little relief from Social Security, Medicare, and Medicaid cost growth.[7]

We now explore two of the possible areas in which nondefense discretionary spending might plausibly increase and so further pressure defense spending.

Natural Disasters

Natural disasters can both reduce federal revenues (by reducing GDP) and increase federal spending. Here we examine two types of disasters: rapidly occurring disasters (such as hurricanes and earthquakes) and slow-moving disasters (such as environmental degradation). Hurricanes Katrina and Rita are examples of rapidly occurring disasters that temporarily interrupted the economy's momentum. They reduced economic growth in the second half of 2005 by about half of a percentage point, in part by pushing up energy prices.[8] Hurricane Katrina alone has cost the federal government over $125 billion to date; for comparison, note that the federal deficit in FY 2006 was $119 billion.

---

[6] Office of Management and Budget, *Historical Tables: Budget of the United States Government, Fiscal Year 2007*, Washington, D.C.: U.S. Government Printing Office, 2006, Table 8.1.

[7] As noted previously, the federal budget equals about 20 percent of the national GDP. Accounting for 18 percent of the federal budget, discretionary spending equals roughly 4 percent of GDP.

[8] Congressional Budget Office, *The Economic and Budget Outlook: Fiscal Years 2006–2015*, Washington, D.C.: Congressional Budget Office, 2005.

The cost of rapidly occurring natural disasters to the federal government is expected to increase because the U.S. population, together with wealth and productivity, are concentrating on the coasts of the continental United States. The West Coast is vulnerable to earthquakes, while the Gulf States and East Coast are vulnerable to hurricanes. The National Climatic Data Center of the National Oceanic and Atmospheric Administration (NOAA) records that there were four hurricanes or tropical storms that resulted in at least $1 billion (in current dollars) in damage to the United States in the 1980s. In the 1990s, there were ten such storms. In the current decade, there have been ten such storms, and more are likely to come. The bands in Figure 2.9 represent storms that have caused at least $1 billion in damage.

Figure 2.9
Cumulative Storm Damage

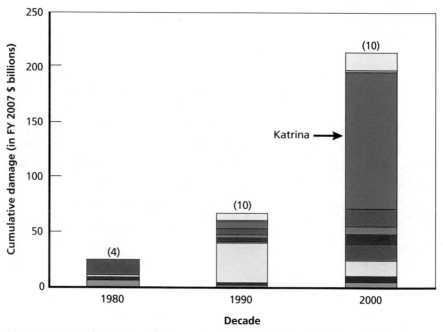

SOURCE: National Oceanic and Atmospheric Administration Satellite and Information Service, "Billion Dollar U.S. Weather Disasters," last updated on July 22, 2008.
RAND MG729-2.9

NOAA's data demonstrate that the frequency of hurricanes or tropical storms that result in at least $1 billion in damage is increasing.[9] Total damage rates are also accelerating. Damage from hurricanes and tropical storms in the 1990s cost more than twice that of storms in the 1980s. Total damage from hurricanes and tropical storms since 2000 has cost more than twice that of storms in the 1990s. Even when the costs of Hurricane Katrina are excluded from calculations, damage from hurricanes and tropical storms since 2000 is quickly approaching the total caused by storms of the entire 1990s—and three hurricane seasons still remain. We therefore conclude that annual disaster-relief costs to the federal government from rapidly occurring disasters will grow.

Global warming is a familiar example of a slow-moving disaster. We chose to examine two less-familiar examples: groundwater depletion and salt contamination of groundwater. Groundwater depletion is caused mainly by sustained groundwater pumping. The negative effects of groundwater depletion are dry wells, reduction of water in streams and lakes, deterioration of water quality, increased pumping costs, and land subsidence. In the remainder of this subsection, we focus on the deterioration of groundwater quality in the United States through salt-water intrusion.

Not all underground water is fresh; in fact, much of the deep groundwater and water in stone beneath oceans is saline. The boundary between freshwater and saltwater is normally stable, but pumping can pull saltwater inland and upward, causing irreversible saltwater intrusion that contaminates the groundwater. When used in irrigation, salt-contaminated water can ruin cropland. At higher levels of concentration, salt renders groundwater undrinkable. Desalination is a feasible but expensive option when groundwater is contaminated by salt.

Desalination in the United States began in Key West in the 1960s. Since then, desalination plants have become larger and more numerous. Texas has 38 desalination plants; California has 33 desali-

---

[9] For a different perspective on the NOAA data, note that from 1980 to 1995, seven storms resulted in at least $1 billion in damage. In the subsequent 13 years, there have been 17 such storms.

nation plants, and plans to build more;[10] Florida has the most desalination plants, with 120 total plants along both coasts. All of these plants purify brackish water. Georgia and the Carolinas, which have experienced depleted groundwater and salt contamination for over a decade, are moving toward desalination. Tampa, Florida, now depends on a large seawater desalination plant to produce 10 percent of its drinking water at a cost of about $150 million annually. California is planning large seawater desalination plants. All of these plants will be expensive and energy-intensive. They will also require that expensive infrastructure, such as new pipelines, be built.[11] Nationwide, billions of dollars will be needed for desalination plant and pipeline construction, and for plant operation and maintenance. At best, plant and pipeline costs will be paid for locally and the water produced will be cheap and plentiful enough to allow any affected farmers to remain competitive. At worst, the federal government will be called on for ongoing relief and, on a regional basis, farmers will become noncompetitive or be forced out of business to conserve water. These worst-case outcomes would increase federal expenses and reduce GDP and federal revenue.

Energy

Crude-oil production in the United States peaked in approximately 1970, producing close to 10 million barrels per day. By 2005, U.S. crude-oil production totaled just 50 percent of its peak level (see Figure 2.10). More positively, the United States remains the world's third-largest oil

---

[10] The California Department of Water Resources reports that California experienced statewide droughts from 1987 to 1992, with a series of dry years in the late 1990s and early 2000s. Southern California experienced a five-year drought from 2000 to 2004. Lakes and reservoirs have been depleted. These events have increasingly drained California's groundwater, with record numbers of wells drilled. A return to normal precipitation patterns would not restore groundwater levels. Large desalination plants are being constructed to alleviate water shortfalls.

[11] Pipelines are needed to bring brackish water or saltwater to desalination plants. Pipelines are also needed to dispose of brine without damaging the environment. See James Miller, *Review of Water Resources and Desalination Technologies*, Sandia National Laboratories, March 2003.

Figure 2.10
U.S. Oil Production, 1900–2005

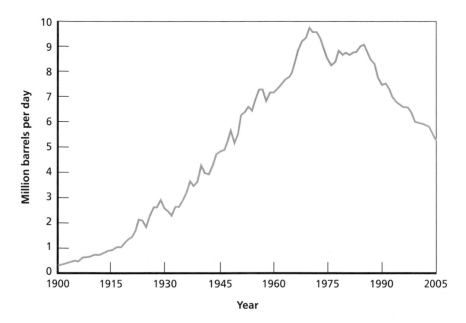

SOURCE: U.S. Government Accountability Office, *Uncertainty About Future Oil Supply Makes It Important to Develop a Strategy for Addressing a Peak and Decline in Oil Production*, GAO-07-283, February 2007.
RAND MG729-2.10

producer, and the U.S. economy is becoming increasingly efficient in its use of oil. From 1983 to 2005, oil consumption in the United States increased at an average annual rate of 1.65 percent, an increase occurring at roughly half the rate of GDP growth. The transportation sector of the U.S. economy, which accounts for about 65 percent of oil consumption in the nation, relies almost entirely on petroleum to operate. Moreover, oil consumption for transportation has increased over recent years despite technological improvements. Oil imports have increased while production has decreased and consumption has increased. As of 2005, the United States imported about 66 percent of its oil and petroleum products. In the near term, the U.S. economy depends on an assured supply of affordable oil. In the long term, continued U.S. economic growth will depend on the nation's ability to innovate in ways that alleviate the problems caused by the fact that increasing global demand for oil is overtaking global oil production.

Globally, the consumption of oil has increased more or less steadily since 1983. The Energy Information Administration (EIA) of the U.S. Department of Energy (DoE) reports that global consumption of petroleum was 84 million barrels per day in 2005; the daily total is expected to reach 118 million barrels by 2030.[12] Much of the additional demand will come from China and India. World oil production has been near capacity in recent years, resulting in recent increases in the cost of oil.

The Government Accountability Office (GAO) analyzed a hypothetical abrupt near-term oil-shortage crisis using a scenario in which Venezuela cuts off its supply of oil to the United States. Using a computer model developed by the DoE, the GAO assessed the impact of a six-month disruption of crude oil resulting in a temporary loss of 2.2 million barrels per day (this daily loss approximates loss rates observed during Venezuela's two-month oil strike in the winter of 2002–2003). The model suggests that such a shortfall would result in a significant increase in crude-oil prices and a reduction of up to $23 billion in the U.S. GDP. In the context of a U.S. GDP that totals more than $10 trillion, this loss represents approximately a 2-percent reduction in the U.S. GDP (see Figure 2.4). The GAO notes that such a disruption would also seriously hurt the economy of Venezuela.[13]

The GAO and the National Petroleum Council (NPC) recently completed two long-term energy studies.[14] The GAO study (1) examined when global oil production might peak, (2) assessed the ability of new transportation technologies to mitigate the peak and decline in oil production, and (3) identified federal agency efforts that could reduce

---

[12] Miller, 2003.

[13] U.S. Government Accountability Office, *Issues Related to Potential Reductions in Venezuelan Oil Production*, GAO-06-68, June 2006.

[14] U.S. Government Accountability Office, *Uncertainty About Future Oil Supply Makes It Important to Develop a Strategy for Addressing a Peak and Decline in Oil Production*, GAO-07-283, February 2007.

uncertainty in the timing of peak oil production or mitigate its consequences. The GAO study reports the following findings:

- Peak oil production cannot be estimated reliably. Unknown factors include the amount of oil remaining globally, the cost and technological feasibility of recovering the oil that remains, and the extent to which oil-producing countries will invest in oil production. The report notes that over 60 percent of all world oil reserves are in countries whose relatively unstable political conditions could constrain oil exploration and production.
- There are considerable tar-sand oil deposits in Canada and there are huge amounts of shale oil in the United States, but it is not clear when or if they will be heavily exploited.
- Alternative fuels (such as corn ethanol) and transportation technologies (such as hybrid vehicles) will displace only the equivalent of 4 percent of the U.S. annual oil consumption by 2015. Under the best conditions, they will displace no more than 34 percent of U.S. consumption by 2025–2030.

The NPC study, which included a remarkable number of diverse participants, reports the following conclusions:

- Coal, oil, and natural gas will remain indispensable to meeting total projected energy demand growth through 2030.
- The world is not running out of energy resources, but the continuing expansion of oil and natural gas production from the conventional sources relied on historically is fraught with risk. These accumulating risks create significant challenges to meeting projected energy demand.
- The concept of energy independence is not realistic in the foreseeable future. However, U.S. energy security can be enhanced by moderating demand, expanding and diversifying domestic energy supplies, and strengthening global energy trade and investment. There can be no U.S. energy security without global energy security.

- Policies aimed at curbing carbon dioxide (CO_2) emissions will alter the energy mix, increase energy-related costs, and require reductions in demand growth.

Defense Spending

Over the past three decades, investment in new weapon systems has been the most variable part of the U.S. defense budget.[15] Under current plans, Army investment spending will peak in 2015 at $42 billion, then decline (most of the additional funds through the peak will be expended on the Future Combat System). Navy investment spending will peak at about $66 billion in 2013, then decline (see Figure 2.11). In

Figure 2.11
Past and Projected Navy and Marine Corps Investment Spending

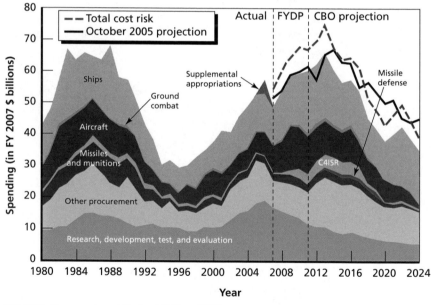

SOURCE: Congressional Budget Office, 2007.
RAND MG729-2.11

---

[15] Daggett, 2005, p. 6.

the Navy, ship acquisition is the main source of budget growth; based on historical cost growth for ships, the CBO calculates that the Navy's investment budget could actually peak at about $75 billion in 2013. The Air Force's annual investment budget will increase from $57 billion in 2007 to about $70 billion between 2012 and 2024. The CBO projects sustained increases in purchases of new tactical aircraft, with production of the F-22 fighter continuing through 2010. Beyond 2010, new programs for light cargo aircraft and new long-range strike aircraft will add to the Air Force investment budget. Joint Strike Fighter costs will also increase after 2010.[16]

Figure 2.12 presents CBO projections of investment and other spending by all the services through 2024. Without considering contingencies and historical cost growth, the CBO projects that DoD budgets

Figure 2.12
Past and Projected Defense Spending

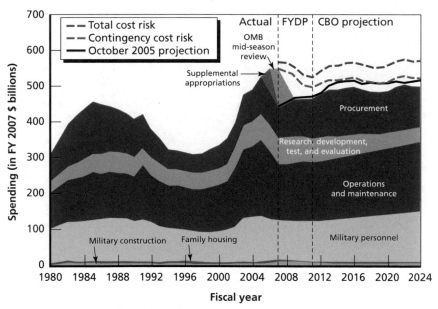

SOURCE: Congressional Budget Office, 2007.
RAND MG729-2.12

---

[16] Congressional Budget Office, 2007.

will grow slightly compared to today's level (in FY 2007 dollars). Cost-risk considerations could increase future DoD budgets significantly.

Conclusions

The following conclusions are supported by data from the Congressional Research Service (CRS), the CBO, The Brookings Institution, the Centers for Disease Control and Prevention, and other organizations:

- Economic growth alone is unlikely to bring the United States' long-term fiscal position into balance. Moreover, issuing ever-larger amounts of debt or dramatically raising taxes could significantly reduce growth. An ever-growing burden of federal debt would have a corrosive and potentially contractionary effect on the budget.[17]
- Unless taxation reaches levels unprecedented in the United States, current spending policies will be financially unsustainable.[18]
- The chief source of pressure on the federal budget will be increased medical costs in Medicaid and Medicare. Structural factors will make it difficult to reduce these costs.
- Under current plans, DoD budgets are projected to increase slightly compared to today's level. In light of cost risks, however, growing procurement costs could significantly increase DoD budgets. With increasing pressure on future DoD budgets, acquisition items with high levels of cost growth could be targeted for reduction or cancellation more frequently in 2020 and beyond.

Note that we expect factors outside the scope of the studies by CRS and CBO, such as trends in natural disasters, to exacerbate pressure on discretionary spending in general and the DoD budget in particular.

---

[17] Congressional Budget Office, 2003, p. 13.

[18] Congressional Budget Office, 2003, p. 13.

The federal budget process is near-term in nature, with budget projections extending only five or ten years into the future. The ten-year "feelers" are now reaching slightly beyond 2017, the first year in which the annual cost of Social Security is expected to exceed tax revenues. This is considered to be a warning event rather than the trigger for real pressure on the discretionary budget. Real pressure on the discretionary budget will come when the budget process officially recognizes the problem.

CHAPTER THREE

The United States' Near Abroad

The United States has been fortunate in that for over a century it has not had to devote considerable military resources to conflicts within the Western Hemisphere. Not since the Spanish-American War of 1898 has the United States made a considerable military effort close to its homeland. The periodic interventions in Central America and the Caribbean from early in the 20th century to the 1994 occupation of Haiti involved relatively small numbers of U.S. forces, and were conducted in the face of negligible opposition. To the north, relations with Canada have been completely nonthreatening since the 1820s. The United States has gained considerable advantage from this situation.

This chapter starts with a review of the current situation in the Western Hemisphere, particularly those nations closest to the United States. We then examine how things might change in the future, emphasizing plausible possible events that could require the U.S. to devote significantly greater resources to ensure its interests in this region.

The Current Situation

The main threats to U.S. interests in the United States' near abroad come from illegal immigration and the inflow of drugs. Both of these challenges come overwhelmingly from south of the United States; Canada is a very minor contributor to either problem. The immigration problem has become an increasingly politicized issue, and could be a major factor in the 2008 presidential election. With an estimated 12 million illegal immigrants in the nation today, the vast majority of whom are from Latin America, how the United States chooses to attack this problem will have major consequences for not only the nation itself but also countries to the south.

The inflow of illegal drugs has been a constant problem since the 1960s. Whereas most illegal narcotics enter Europe from Central Asia, the primary sources of drugs in the United States are Central and South America. The sheer scale of the drug trade has had a corrupting influence on many Latin American nations. In the case of Colombia, for example, the main leftist insurgent groups—the *Fuerzas Armadas Revolucionarias de Colombia* [Revolutionary Armed Forces of Colombia] (FARC) and the *Ejército de Liberación Nacional* [National Liberation Army] (ELN)—have in recent years increasingly become criminally oriented organizations interested in making money from the drug trade as opposed to politically focused opposition movements. Many of the Caribbean nations are also deeply involved in the flow of illegal drugs into the United States. A phenomenon similar to the protracted struggles in Colombia has recently emerged in Mexico. The new Calderon government has launched a major crackdown against a variety of very violent drug cartels that are operating in western and northwestern Mexico. Apparently, a number of the drug cartels have been able to acquire significant amounts of small arms from the United States. Unlike the FARC and ELN, the Mexican drug cartels have no ideological pretensions. Rather, they behave like the sophisticated organized crime groups of the Medellin and Cali cartels in Colombia. To date, the U.S. military has supported other government agencies that are leading the fight against the drug trade. The next American administration may be asked by the Mexican government for outright assistance in the form of either military hardware (such as helicopters and aerial sensors) or more-covert intelligence community support of the kind that the United States supplies to Colombia.

An important recent trend is the resurgence of anti-American, leftist political movements in Central and northern South America. The election—and reelection—of Hugo Chavez in Venezuela is the most obvious example of this trend. Chavez is ideologically very closely linked to Cuba's Fidel Castro (indeed, they are close personal friends) and sees himself as a kind of new Bolivarian figure in the region, championing the causes of populism, anti-Americanism, and a socialist political-economic model. In recent years, Chavez has consolidated his position within Venezuela by eliminating political opponents, purging the army and police, and closing the radio and television stations of the opposition. Currently, Chavez is still focusing on tightening his control within Venezuela. Once that is done, however, he could be motivated to increasingly interfere with the internal affairs of other nations—such as neighboring Colombia, long threatened with a leftist insurgency.

Future Trends and Possibilities in Northern South America

Tension may arise between the Andean states that side with Venezuela (e.g., Bolivia and Ecuador) and those that reject Chavez's ideals (e.g., Peru and Colombia). Currently under the leadership of President Lula, Brazil constitutes a "five hundred pound regional political and economic gorilla" that will likely become a critical swing vote in this regional contest for influence. Already there is evidence that Chavez may have overreached himself by attempting to join Mercosur and dominate it by positioning Venezuela as the major supplier of gas and oil to the South Cone. Chavez's public outbursts against Brazil's massive and successful conversion of sugarcane to ethanol may be part of a political campaign to shore up Venezuela's role as the energy supplier of choice. Even if he faces setbacks, however, Chavez will continue to try to forge a strong regional alliance that takes on a systematic anti-American orientation. Although some dismiss it as a stunt of little consequence, Chavez's recent economic and security overtures to Iran strongly suggest that the next U.S. administration will have to pay closer attention to the political, economic, and (possibly) security dimensions of Latin America.

In addition to the possibility of a new wave of leftist insurgencies in the region, there is virtual certainty that the flow of illegal narcotics from Central and northern South America will continue through the medium and probably long term. The drug economy has become so entrenched in the region, and has co-opted so many of the region's national governments, that the situation will probably not improve much in the foreseeable future. Indeed, the United States' likely continued demand for illegal drugs will help ensure that there will be a robust flow of narcotics from Central and South America into the United States. This reality guarantees some level of U.S. military commitment to the war on drugs in the region, whether or not the threat from leftist insurgencies increases.

Future Trends and Possibilities in the Caribbean

The biggest near-term variable in the Caribbean is the future of Cuba after Fidel Castro passes from the scene. The Cuban economy has been very weak for decades, and Castro's continued adherence to Marxist ideology has contributed to the nation's isolation and economic underperformance. Fidel's brother Raoul, long the key figure overseeing Cuba's security services, has become the successor to power in Havana. However, Raoul is 75 years old at the time of this writing, and therefore, even if he does succeed his older brother, his will be a transitional period. Less clear is who might emerge as the long-term successor to Fidel.

Cuba's economy provides a modest if not meager existence for the vast majority of the country's population. Calls for significant economic liberalization might lead to a serious power struggle between reformers and traditional hardliners. The Cuban Communist Party reformers, with Raoul Castro's overt or covert leadership, will most likely attempt to move Cuba toward a variant of China's authoritarian-state capitalism. Cuba is the second-largest source of sugar production in the Western Hemisphere. With Brazilian assistance, and contrary to Fidel's diatribes against the conversion of corn to ethanol, a significant portion of that production could help alleviate Cuba's dependence on foreign sources of oil. When coupled with major Cuban investments in wind and solar power, this new source of energy might give the Communist Party reformers a political edge that could ensure their political dominance even after the departure of both Castro brothers. On the other hand, Fidel may outlast his younger brother, who is in ill health, and paralyze any meaningful economic reform until his passing. Then, the stage will be set for the prospect of a post-Castro Cuba that degenerates into semi-chaos or even civil war.

In the event of such a crisis, the United States might feel powerful pressure to intervene militarily, or at least to provide considerable support to one or more factions fighting for power in post-Castro Cuba. Should the United States intervene directly, the consequences could include a significant, multiyear commitment of considerable numbers of U.S. troops who would be trying to stabilize a country of over 11 million people, many of whom would be hostile to the United States. Obviously, any unilateral U.S. intervention without the sanction of the United Nations (UN) and the Organization of American States would be highly controversial.

In the remainder of the Caribbean, the United States will continue to face the challenge posed by the inflow of illegal drugs from economically poor nations. This will probably continue to be a problem through 2025, since there is no indication that there will be any major breakthrough on the counternarcotics front. In the medium term, increased development of sugar-based energy (ethanol), wind and solar power, and oil drilling in the Caribbean will lead to new economic opportunities for these countries. Whether these new energy-sector opportunities decrease the importance of narcotics as a means of income in those nations is questionable. After all, Colombia's overall economic performance has been robust from time to time, but this success has not led to any reduction of the illegal drug trade that continues to be driven by demand from North America and Europe. One economic bright star in the region is Panama, which is enjoying an economic boom brought on by an improved and somewhat less-corrupt government, the country's emergence as a major tourist and service center, and major investment in the modernization and expansion of the Panama Canal.

Of special interest to the United States will be the evolution of domestic politics in the Caribbean and northern South America. The tension between generally pro-American Colombia and the nuevo-Marxist Venezuela of Hugo Chavez shows the contrasting visions that are at work in the region. Whether pro-American leaders and parties prevail in the coming years or whether Chavez represents a lasting wave of leftist anti-Americanism remains to be seen. How the leaders of the region organize their economies, deal with the aspirations of their peoples, and handle their own relationships to powerful groups in their own countries (such as the security forces) will be the key variables as the region moves into the second decade of the 21st century and beyond.

An Overview of Future Trends in the Region

As mentioned earlier, there is now a rise of nuevo-leftist attitudes and regimes in the region; a primary example is the regime of Hugo Chavez in Venezuela. The United States will face a significant challenge in the near and medium terms if Chavez elects to provide increasing support to the FARC and ELN insurgents in Colombia. In the recent past, President Uribe of Colombia made important strides in putting his country on a more stable path, but the Colombian government remains vulnerable to increased insurgent attacks. Should Chavez decide to dramatically increase his support to the FARC and ELN, the situation in Colombia could deteriorate dramatically. Indeed, using Venezuela's considerable oil revenue to finance such a move, Chavez could elect to stir up trouble in various parts of Central America as well. To date, the United States has been less concerned about the rise of anti-American leftist regimes in the region than it was during the Cold War. This is probably due to the fact that the new anti-American regimes lack a powerful external backer.

So far, the level of U.S. commitment to friendly governments in Central and South America has remained modest. In the event of a new wave of leftist insurgencies,[1] however, the United States might have to become much more seriously engaged in the region. A very important variable in this situation is whether Brazil, Argentina, Peru, and Chile choose to stand aside, support the United States, or support the various leftist movements.

---

[1]   Potentially sponsored by Venezuela and Cuba, for example.

The Special Case of Mexico

By far the most important strategic question for the United States in its near abroad is "whither Mexico?" The majority of the illegal immigrants in the United States come from Mexico. Unlike many other Latin American groups, a large portion of the Mexican immigrant population does not assimilate into U.S. culture, remaining instead a separate, Spanish-speaking subgroup with limited economic opportunities. The magnitude of the Mexican immigrant challenge has been a growing source of tension between the United States and Mexico for years, and this tension is not likely to dissipate significantly through at least 2017.

How the United States and Mexico resolve the immigrant problem—or not—has major implications for both nations. Should the situation remain unresolved, tension between the two countries will likely increase. The issue could become increasingly polarizing for the U.S. population, and relations between the two countries could be increasingly strained. The success or failure of Mexico's domestic economy and political system will have a major bearing on the immigrant issue.

Mexico's economy has benefited in many ways since the North American Free Trade Agreement was signed during the Clinton administration. However, there have been major "winners" and "losers" within Mexico, as there have been in the United States. Mexico's economy remains fragile and is still underperforming in relation to the size of the country's population. If Mexico's economy does not grow to meet the needs of an expanding population, domestic problems within that nation will increase over time. Continued high levels of corruption within Mexico do not help in this regard.

The narrowly elected Calderon government has shown considerable political skill in building working political coalitions with its nationalist rivals, including the *Partido Revolucionario Institucional* [Institutional Revolutionary Party] (PRI), and has effected significant financial and economic reform by causing members of the leftist *Partido de la Revolución Democrática* [Party of the Democratic Revolution] (PRD) to defect. Unlike the disappointing era of President Vicente

Fox, the Calderon government may set Mexico on a path of sustained and robust economic growth that addresses some of the social ills of the society.

On the other hand, these attempted financial and economic reforms may fail, and there is a real possibility of significant economic and political instability in Mexico, at least through the medium term. The tension following Mexico's 2005 elections, whose results were seriously challenged for weeks by Mexico City's mayor, shows that the political process in the country remains fragile. As noted above, the continued influence of rich drug cartels that have undermined the nation's police and courts through extensive bribery is another factor contributing to political instability.

A low-probability but worst-case scenario for the United States in the medium term is the possible emergence of an extreme, anti-American, nationalistic political movement in Mexico. Such a movement would likely grow out of a political renaissance of the left-of-center PRD party following a politically and economically failed Calderon regime. If those Mexican political events are joined by a continuing and major illegal immigration problem, relations between the United States and Mexico could deteriorate to dangerous levels. In this situation, the U.S. military, and particularly the Army and the National Guard, might have to focus on the south to an extent not seen in over a century.

CHAPTER FOUR

Strategic Trends in the People's Republic of China

China is living out a Faustian bargain. Its growth-at-any-cost policy has provided three decades of robust economic growth and has increasingly urbanized its once largely peasant population. The People's Liberation Army (PLA) is transforming from a huge defensive ground force with antiquated equipment into a modern military capable of limited power projection. In exchange for wealth and might, however, China has sacrificed its environment, its natural resources, and the well-being of its elderly and rural residents. Unhealthy air and undrinkable water are common. Cropland is turning into desert. Energy and water resources are being depleted. Many elderly citizens lack retirement incomes, and access to health care is increasingly scarce. All of these problems are worsening as environmental degradation is largely ignored and the population ages.

In the next 10–15 years, China will confront the complex and intertwined problems of dwindling energy resources, increasing energy needs, pollution, more-expensive and increasingly scarce raw materials, slowing economic growth, and the emerging unmet needs of its elderly population. These forces may collectively discourage Chinese military development and deter China from engaging the United States militarily.

The Current Situation

The PRC has enjoyed nearly 30 years of stellar economic growth that has enabled it to undertake significant military modernization in a remarkably short period. Even if growth slows, we project that China's GDP in 2020 is likely to total $7 trillion to $8 trillion; other fore-casters believe it could reach $15 trillion—about 8 percent more than the United States' 2007 GDP. Whatever the actual figure, the growth of China's economy will support near- to medium-term expansion of China's military capabilities. Therefore we argue that mounting inter-nal pressures, rather than economic limitations, may constrain China's ability to expand military capabilities in this period.

Thanks to decades of almost-constant economic expansion, China's once largely peasant population is becoming progressively urban-ized, its previously agrarian economy is largely industrialized, and many of its coastal cities are increasingly modernized. The military is transforming itself from a huge defensive ground force with antiquated equipment into a modern military capable of limited power projection. However, in exchange for these near-term gains in wealth and power, China has sacrificed its long-term energy resources (primarily coal), embarked on an unsustainable economic-growth path, skimped on pensions and health care for its elderly, and accepted extensive environ-mental damage that will impede economic development and require remediation. These problems are tightly intertwined, making them harder to address. For example, as its hard-coal reserves are depleted, China must increase its dependence on lignite, the most polluting form of coal with the lowest energy content; this shift will worsen air pol-lution and its attendant health risks and increase the incidence of acid rain.

Hence, to understand China's future role in the world, one needs to look first at China's internal changes and challenges. The pressures and passions arising from China's domestic conditions will profoundly shape China's role as a global actor. Chinese leaders often discuss "strengthening comprehensive national power," which means develop-ing all aspects of national power—political, economic, and military. This chapter takes a similarly broad view of the factors that are likely to influence future development trajectories in China. It also considers how these factors could affect Beijing's ability to maintain a high level of defense expenditures once other domestic interests begin clamoring for greater attention and resources.

The next two decades will likely see increasing turbulence in China as the Chinese Communist Party (CCP) struggles to maintain single-party rule and sustain economic growth while adapting government spending and policies to the needs of China's rapidly changing society. The size and complexity of the domestic challenges facing China in the near to medium term almost certainly preclude the possibility of a smooth-glide trajectory to superpower status. By 2030, China will have to cope with the fact that it has become the first nation to grow old before it grows rich. China's growth-at-any-cost approach to economic development has already wreaked havoc on its natural environment: Desertification and pollution will likely threaten China's water supply within a matter of years. Although China's comprehensive national power is likely to grow, the pace of growth will slow significantly as latent domestic pressures surface. China should be able to maintain its military-modernization path through roughly 2015, but beyond that time it may become too distracted by internal concerns to continue its rapid military buildup.

China's Energy Future

Consumption Pattern and Energy Mix

China's energy profile—its overall consumption pattern and energy balance—reflects the path that the country's economic development has taken as well as the relative strength (or weakness) of the state institutions that govern this growth. Having tripled since 1980, China's total primary energy consumption (TPEC) has surpassed the consumption levels of leading Western economies and is rapidly approaching U.S. levels (Figure 4.1). In 2005, China's GDP totaled 19 percent of the U.S. GDP at market exchange rates and 43 percent as measured by purchasing-power exchange rates, but its TPEC was slightly more than half the U.S. level. The International Energy Agency (IEA) estimates that China's primary energy demand will exceed that of the United States by 2030, constituting roughly 20 percent of the global primary energy demand and making China the world's largest energy consumer.

Figure 4.1
China's Primary Energy Demand

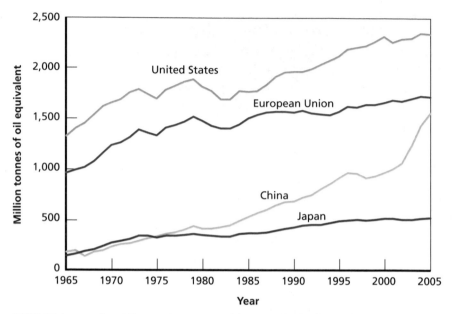

SOURCE: International Energy Agency, *World Energy Outlook 2006*, Paris, 2006.
RAND *MG729-4.1*

China is expected to account for more than 30 percent of the projected increase in global energy demand between now and 2030.[1]

Chinese energy consumption per capita is still comparatively low, totaling roughly one-eighth of U.S. consumption and one-third of European consumption levels.[2] One positive trend is that Chinese energy intensity—the amount of energy consumed per unit of GDP—dramatically decreased between 1978 and 2000. In this period, the Chinese economy expanded at almost 10 percent per year in real terms, but energy demand grew more slowly. As a result, by 2000, the Chinese economy consumed only a third of the energy per unit of GDP as it

---

[1]   International Energy Agency, *World Energy Outlook 2004*, Paris, 2004.

[2]   Energy Information Administration, *International Energy Annual 2005*, June–October 2007b.

did in 1978. China's success at improving energy intensity on such a scale is impressive.

Since 2001, however, progress on reducing energy intensity has slowed. Recent data suggest that China's energy demand elasticity (the ratio of growth in energy demand to GDP growth) tripled from a pre-2001 average of 0.5 to more than 1.5 in 2006.[3] In other words, a 1.0-percent increase in GDP now requires over 1.5-percent growth in energy demand. This suggests that gains in energy efficiency in certain sectors of the Chinese economy have been offset by the rapid expansion of other energy-intensive sectors.[4] China remains an inefficient user of energy: To produce $1 of GDP, the PRC requires three times more energy than the world average, 4.7 times more than the United States, 7.7 times more than Germany, and 11.5 times more than Japan.[5]

China now covers about 70 percent of its TPEC needs through coal; oil and natural gas are the second and third most popular sources of energy (Figure 4.2). China reports that it consumed 2.38 billion metric tons (tonnes) of coal in 2006, roughly twice the U.S. consumption level in the same year.[6] Coal will remain the dominant source of China's energy for the near term; oil's share of TPEC will increase sharply, and growth in gas consumption will be far more modest. What appears certain is that as the years go by, China will have to import a larger percentage of its coal to meet its requirements. This greater dependence on foreign coal could affect international coal prices (however, because coal is more globally plentiful than oil, increased Chinese coal imports are not likely to affect coal prices). Greater dependence will probably result in increasing Chinese ties with the nations from which it imports coal (e.g., Australia). To improve its coal infrastructure, China may also have to improve the ports that receive coal imports.

---

[3] National Bureau of Statistics of China, *China Statistical Yearbook*, Beijing: China State Statistical Press, 2005, p. 147.

[4] Nicholas Lardy, "China: Rebalancing Economic Growth," Peterson Institute for International Economics, 2007.

[5] Doug Ogden, *China's Energy Challenge*, The Energy Foundation, 2004, p. 5.

[6] British Petroleum, *World Statistical Yearbook 2006*, London, 2006b.

Figure 4.2
China's Projected Primary Energy Mix

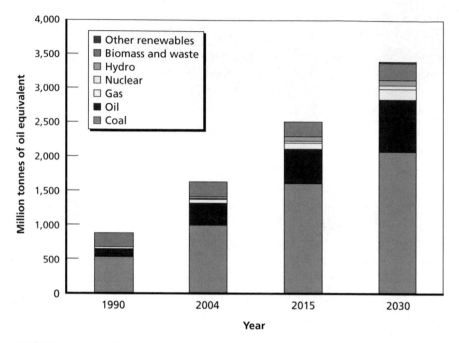

SOURCE: Ogden, 2004.
RAND MG729-4.2

Until recently, China was self-sufficient in oil; in fact, China was actually an oil exporter until 1993. Oil production in China has not yet peaked, but China's demand for oil under a burgeoning economy has outstripped production (see Figure 4.3). China's urbanization, together with the increased affordability of automobiles, has accelerated and will continue to accelerate China's oil demand. At present growth rates, China will be both the world's second-largest oil consumer (behind the United States and before the European Union [EU]) and the world's second-largest oil importer in 2030. Its overall import share will rise from 46 percent in 2004 to almost 80 percent by 2030.[7] If EU member states are considered individually rather than collectively, China now is the world's second-largest consumer of petroleum products.

---

[7] International Energy Agency, 2006, p. 101.

Figure 4.3
China's Oil Balance

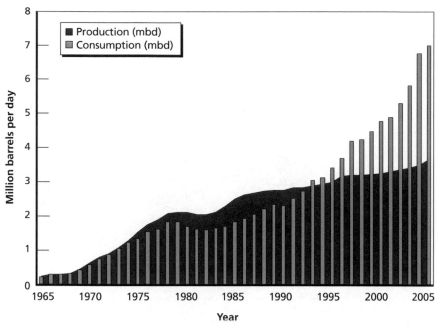

SOURCE: Energy Information Administration, 2006a.
RAND MG729-4.3

Chinese urbanization and the increased affordability of auto-mobiles have accelerated and will continue to accelerate China's oil demand. As a newcomer to the international oil market, China first turned to lesser producers (such as Oman) and sought exclusive drill-ing rights as a means to secure oil supplies. More recently, China has turned to Africa and Iran for oil. Among African countries, Angola emerged as China's largest and most important oil supplier in 2006.[8] Perhaps deliberately, China has, through its choice of oil suppliers, avoided competition or confrontation with the United States in the international oil market (see Figure 4.4). China now draws its 3.6 mil-lion barrels per day of crude imports from almost every major oil-producing region. Although Venezuela sells a relatively small share of

---

[8] Energy Information Administration, *Country Analysis Brief: China*, August 2006a.

Figure 4.4
China's Oil Import Sources

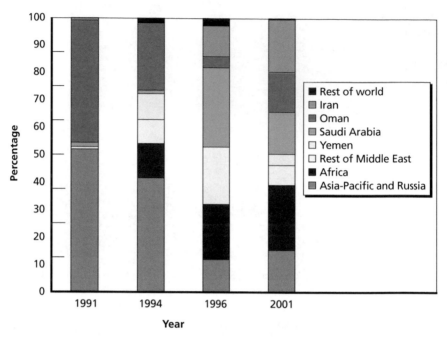

SOURCE: Energy Information Administration, 2006a.
RAND *MG729-4.4*

its total production to China, Caracas has announced plans to significantly expand oil exports to China by the end of the decade and is thus
likely to grow as a source of China's oil.

Despite Chinese efforts to secure African energy resources and
flank them with diplomatic and financial side payments, the Greater
Middle East (GME) has become China's single most important source
of oil. At present, it provides more than 40 percent of Chinese oil
imports.[9] China's import dependence on the GME will gain further
momentum because the vast majority of global oil reserves is concen-
trated in the Persian Gulf region—a fact that cannot be circumvented
by Chinese "energy diplomacy" in other parts of the world.

---

[9]   Energy Information Administration, 2006a.

Although the central Chinese government has attempted to increase production of hydropower and other nonfossil fuels, these sources' overall contribution to China's total energy supply will remain comparatively low. Nuclear power, though growing, will not be able to satisfy China's growing demand for energy.

The strongest drivers of current and future Chinese energy-demand growth are power generation, heat plants, and heavy industry and residential consumption. On average, one coal-fired power plant has come on line in China every week since 2005.[10] Transportation remains less important and is expected to represent only 10 percent of China's TPEC in 2030.

China's Coal Industry

China depends on coal for about 70 percent of its energy. It has the world's third-largest coal reserve (behind the United States and Russia) but is the world's largest producer of coal. Chinese coal production has grown at about 10 percent per year since 2000 and has increased more than 2,000 percent since 1992 (see Figure 4.5). Such growth cannot be sustained indefinitely, but there is uncertainty and disagreement about when China's coal production will peak and when China will exhaust its coal reserves. China will have to explore and develop new fields and import more coal to cover future energy demands. Recent analyses suggest that China's coal production will peak sometime between 2015 and 2025, with coal production levels between 2030 and 2040 falling below current levels unless significant new reserves are found. Coal shortages may cause power outages, inflation, and plant shutdowns in China.

Note that even if China does uncover significant new coal reserves, the following issues could arise: difficult and costly extraction, poor coal quality (in terms of its energy content), and infrastructure challenges. Appendix B describes several recent analyses of China's coal future and offers direct supporting evidence for our findings, includ-

---

[10] Woodrow Wilson Center for International Scholars, "Energy in China Fact Sheet," 2005.

Figure 4.5
China's Coal Production Since 1970

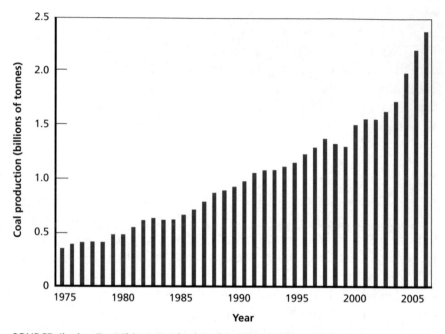

SOURCE: Jianjun Tu, "China's Botched Coal Statistics," *China Brief*, Vol. 6, No. 21, October 25, 2006.
RAND *MG729-4.5*

ing evidence that China recently became, for the first time, a net coal importer.

Macroeconomic Trends

For two and a half decades, China's unfolding economic success story has awed and fascinated the international community. China has grown very rapidly since 1978, posting an average growth rate of 10 percent (according to official figures).[11] In 2005, China's GDP reached

---

[11] A number of economists question the reliability of China's official economic statistics. Some economists argue, for instance, that official numbers could overstate actual growth.

$2.225 trillion, surpassing the United Kingdom and France to make China the world's fourth-largest economy.[12] In the same year, China also outpaced Japan and became the world's third-largest exporting nation.[13] In all, Chinese real per capita output in 2005 was nine times greater than in 1978 at the beginning of the era of reforms.

Although the demonstrated resilience of China's economic growth is remarkable, it is not unprecedented. Other Asian economies, such as those of Taiwan, South Korea, and Japan, demonstrated similarly strong growth at the peak of their development trajectories; most of these economies then experienced subsequent slowdown to more-sustainable levels. Assuming that China's economy continues to grow over the next two decades, it is highly likely that it too will experience a deceleration in its growth rate. Unlike its Asian predecessors, China's path is further complicated by the unique challenges that it faces as the world's most populous nation. The sheer scale of China's internal needs is daunting. Since 1990, an estimated 400 million Chinese have been lifted out of absolute poverty; yet another 415 million still live on less than $2 per day.[14] In absolute terms, China's GDP in recent years has been impressive, but when viewed in per capita terms, the considerable distance between China and most advanced developed nations is clear. In 2005, China's GDP per capita was roughly $1,700, a level comparable to U.S. per capita GDP in approximately 1850.[15]

Finally, China's future is harder to discern because of the unpredictability associated with its status as both a developing and transitioning economy. China has not completed its transition from

---

See Lester Thurow, "A Chinese Century? Maybe It's the Next One," *New York Times*, August 19, 2007.

[12] Peter S. Goodman, "Too Fast in China? Stunning Growth May Have a Built-In Problem: Overcapacity," *The Washington Post*, January 26, 2006, p. D01.

[13] Frances Williams, "China Exports Overtake Japan," *Financial Times* (London), April 14, 2005.

[14] C. Fred Bergsten, Bates Gill, Nicholas R. Lardy, and Derek J. Mitchell, *China: The Balance Sheet*, New York: CSIS & Institute for International Economics, Public Affairs, 2006, p. 18.

[15] Bergsten et al., 2006, pp. 4, 19.

the closed, communist command system of the late 1970s to an open economy governed by market forces. Deeper reform will require the CCP to cede more control to the private sector, thereby weakening the government's role in the economy. Thus, the critical economic policy question is whether the ruling party will continue to implement deeper economic reforms despite knowing the political risk that such actions could entail.

GDP Forecasts

Forecasts of China's GDP rates through 2020–2025 typically predict that average growth will range from 5 percent to 7 percent. Forecasts of China's GDP for the same period vary more dramatically depending on whether GDP is calculated at market exchange rates or using purchasing-power parity (PPP).[16] In March 2006, Chinese Premier Wen Jiabao delivered the "11th Five-Year Plan (2006–2010)," which projects GDP growth of 7.5 percent between 2005 and 2010 and aims to raise per capita GDP from renminbi (RMB) 13,985 to RMB19,270 in the same period. This reflects the current leadership's goal of raising the absolute level of GDP in 2020 to four times the level in 2000.[17]

China is well on its way to meeting and surpassing that goal. In 2005, the DoE estimated that China's GDP growth rate would average 7.2 percent through 2025. In 2006, however, Deutsche Bank predicted average growth in China to register 5.2 percent through 2020.[18] The *Economist* Intelligence Unit forecasts 6.76-percent average growth

---

[16] Although this monograph presents both market exchange and PPP estimates, we maintain that PPP rates alone distort economic reality. While it is true that market exchange rates fail to fully appreciate Chinese domestic buying power (due to an undervalued currency and low domestic prices), PPP rates inflate the future size of China's economy relative to that of the United States. As China grows, inflation will ensure that consumer prices rise to levels more comparable with those in the United States. Thus, the gap between market exchange and PPP rates will narrow by 2025. Moreover, global markets determine prices for many of the commodities that matter most for the sake of comparison.

[17] Richard McGregor, "China to Drop Rigid Ambitions for Growth," *Financial Times* (London), March 7, 2006.

[18] Stephan Bergheim, "Global Growth Centres 2020: Challenges and Choice for European Policymakers," *Deutsche Bank Research*, November 2006.

for China through 2020, with growth expected to start slowing from 10.3 percent in 2007 to 9.4 percent in 2008.[19] In a 2005 study, a major multinational petroleum firm predicted that average growth in China will vary between 6.7 percent and 8.4 percent per year through 2025, with variations depending on global economic conditions and the internal Chinese political environment.[20]

Using a hybrid approach that harnesses the advantages of both market exchange and PPP rates, RAND forecasts China's GDP to grow by an average annual rate of 6.59 percent through 2020.[21] Richard N. Cooper, an economist at Harvard, provides a useful and more detailed growth scenario that allows for focus on what this all means for China's development trajectory, relative to the United States, through 2025. Using the DoE forecast of 7.2-percent annual growth, and assuming 1-percent per year appreciation of the RMB relative to the U.S. dollar, China's GDP in 2025 would amount to about $7.4 trillion (as measured in FY 2005 dollars. During the same period, if the U.S. economy grows at 3 percent a year, China's GDP will total roughly one-third the size of the contemporary U.S. economy, but will have reached the size of the U.S. economy in 1988. Chinese per capita GDP will remain at only about one-twelfth the level enjoyed in the United States and Japan, despite having increased fivefold since 2000.[22] With an economy the size of the U.S. economy in the late Cold War, China's future leadership will have the resources and capabilities it needs to project power internationally through foreign aid, defense expenditures, and other means.

---

[19] *Economist* Intelligence Unit, "Country Report: China," June 2007b, p. 12.

[20] Jeroen van der Veer, "Shell Global Scenarios to 2025," The Hague: Royal Dutch/Shell Group, June 2005.

[21] For an explanation of this methodology, see Keith Crane, Roger Cliff, Evan Medeiros, James Mulvenon, and William Overholt, "Modernizing China's Military: Economic Opportunities and Constraints," unpublished RAND Corporation research, 2005, pp. 44–49.

[22] Crane et al., unpublished RAND Corporation research, pp. 44–49.

Obstacles to Future Growth

Although in the near term China can probably continue to exploit some of those factors that have driven its rapid growth thus far, including greater openness to trade, improved technology, and a large labor pool, many of these factors will contribute less to growth than in the early years of China's economic takeoff. Furthermore, structural impediments and habitual problems in China's economic system could conspire to weaken future economic growth.

State-Owned Enterprises and Banking Reform. Most economists agree that the biggest threat to future growth in China, and the most fundamental structural issue within its economy, is the continued existence of state-owned enterprises (SOEs) that suck resources away from enterprises that may be more efficient and profitable.[23] At present, approximately 150,000 SOEs are in operation. Beijing has privatized or partially privatized a number of SOEs, some of which have become profitable. Yet most SOEs are kept solvent by loans from state-owned banks for political reasons—namely, the CCP fears the specter of unrest caused by further massive layoffs, and it is wary of crossing the ideological threshold that full privatization would represent.

The Organisation for Economic Co-operation and Development (OECD) estimates that 7 percent of Chinese companies have negative equity or negative rates of return. State-controlled companies employ 11 percent of all Chinese workers and tie up 23 percent of assets and 22 percent of outstanding debt. When state-controlled companies with sub-par (<5 percent) rates of return are factored in, 10 percent of Chinese firms have negative equity or negative rates of return. These companies employ 20 percent of Chinese workers and tie up 20 percent of assets and 40 percent of outstanding debt.[24] Collectively, private and foreign-invested enterprises generate 52 percent of China's GDP, yet they account for only 27 percent of total bank loans.[25] State-owned

---

[23] See Ira Kalish, *Global Economic Outlook 2007*, Cambridge, Mass.: Deloitte Research, 2007, pp. 4–5.

[24] Organisation for Economic Co-operation and Development, *OECD Economic Surveys: China*, Vol. 2005, No. 13, September 2005b, p. 104.

[25] Kalish, 2007, p. 5.

banks funnel the remaining three-quarters of bank loans into the state sector. Consequently, the private sector—the real engine of growth and innovation—is hampered by a system that habitually supports the state sector for political purposes.

This dysfunctional relationship between the SOEs and the banks makes thorough reform of the SOEs a prerequisite to restructuring China's frail financial system. The frailty of the financial system derives in large part from the nonperforming loans (NPLs) that are created through these thinly veiled subsidies to SOEs. In January 2001, after years of public denial, the Bank of China officially estimated that NPLs account for about 25 percent of all outstanding loans. This figure totaled roughly one-quarter of China's GDP at the time, but independent economists felt certain the actual NPL ratio was far higher, with NPLs potentially accounting for 60 percent of Chinese GDP as of 2002.[26] A high NPL ratio causes an enormous drain on the productive use of capital and translates into a very low average rate of return.

China has managed banking problems by periodically pumping money into the state-owned commercial banks to keep them afloat and by transferring their bad loans to state-owned asset-management companies. In other words, the debt was simply shifted between arms of the government. By the end of 2005, the asset-management companies claimed to have disposed of two-thirds of the NPLs, but without much supporting evidence, most economists doubt this claim. Even if the claim were true, large volumes of new NPLs will continue to be created as long as the "flow" problem goes unaddressed. With nearly three-fourths of all bank loans still flowing to unprofitable SOEs, NPLs will inevitably resurface as a challenge to the solvency of China's major financial institutions. Furthermore, the government has yet to fashion a workable solution to the solvency problems that plague rural credit cooperatives. These institutions collectively hold a quarter of the bank-

---

[26] "China Bank Bailout Could Need \$290 Bln: Report," *People's Daily Online*, January 27, 2003.

ing systems' total assets and are believed to possess at least $100 billion in unresolved bad loans.[27]

In addition to obstructing the emergence of an efficient capital-allocation mechanism, the state's presence lingers in other ways throughout business in China. Most large enterprises in China have a party committee and "the power of Party committees (especially in personnel matters) has become detrimental to the performance of listed enterprises."[28] Central government interference also continues to distort the market in sectors such as energy and credit.

Dependence on Exports, Investment, and Undervalued Currency as Drivers of Growth. Over the past five years, exports have increasingly contributed to growth in China's extremely trade-dependent economy. In 2005, China's total trade in goods (exports plus imports) was equivalent to 64 percent of GDP—a rate higher than found in most other continent-sized economies.[29] Exports as a percentage of China's GDP rose from 25 percent in 2002 to approximately 39 percent in 2006. According to the *Economist* Intelligence Unit, net exports accounted for one-quarter of China's real GDP growth in 2005–2006. From this growing dependence on trade, the *Economist* Intelligence Unit concludes that "if export markets sneeze, China may now catch a cold."[30] China is thus vulnerable to fluctuations in export demand. Given that the United States absorbs fully one-fifth of China's total exports, if existing trade frictions get out of hand or if economic difficulties emerge in the U.S. economy, China's GDP will suffer.

China is also heavily dependent on investment-driven growth, some of which is foreign in origin. Since 2001, overall investment in China has grown much more rapidly than GDP. By 2004, the share of GDP derived from investment reached 40 percent.[31] Investment

---

[27] Barry Naughton, *The Chinese Economy: Transitions and Growth*, Cambridge, Mass.: MIT Press, 2007, p. 467.

[28] Organisation for Economic Co-operation and Development, September 2005b, p. 111.

[29] Naughton, 2007, p. 377.

[30] *Economist* Intelligence Unit, "China Economy: Critical Issues—Export Risks," June 22, 2007c.

[31] Bergsten et al., 2006, p. 26.

in certain sectors—particularly steel, aluminum, cement, real estate, and electric power—has achieved such high levels in recent years that supply has greatly outpaced demand; this has given rise to concerns that the economy could overheat. Most economists believe that stronger domestic consumption is the best solution to China's dependence on trade and investment for growth. However, until Beijing takes steps to strengthen and expand the nation's social security net and improve access to free health care and education, private consumption will continue to grow more slowly than incomes as Chinese citizens save to guard against illness and old age.

The Chinese government's efforts to prevent the appreciation of its currency against the U.S. dollar are a prominent source of trade tension between the United States and China. This devaluation is also widely cited by economists as an impediment to China's sustainable growth and financial stability.[32] Economists believe that the RMB is devalued by at least 15–30 percent. Beijing's policy of distorting the RMB's value in global currency markets has contributed to overproduction in certain sectors of the economy, fueled trade disputes, and, in the last year, given rise to domestic inflationary pressures. The low value of the RMB is also an important factor in China's low per capita GDP: The undervalued currency impoverishes Chinese citizens by reducing their consumption power on the global market.

Widespread Corruption. Corruption is the most serious, chronic threat to the CCP's legitimacy and to the overall administrative capacity of the Chinese state.[33] The popular perception that corruption within the government is almost universal severely undermines the Party's ideological claim that it protects and represents the interests of China's working class. A 2003 survey conducted by the Chinese Academy of Social Sciences (CASS) reveals that urban residents from virtually every locality, every income level, and every educational background believed that the primary beneficiaries of 25 years of economic growth

---

[32] Morris Goldstein and Nicholas Lardy, "A Modest Proposal for China's Renminbi," *Financial Times* (London), August 26, 2003.

[33] Melanie Manion, *Corruption by Design: Building Clean Government in Mainland China and Hong Kong*, Cambridge, Mass.: Harvard University Press, 2004.

have been the Party and government cadres.[34] Another CASS survey from 2003 finds that 55 percent of urban residents surveyed ranked corruption as one of China's two most serious social problems—second only to unemployment and layoffs.[35]

Reports from foreign investors and nongovernmental organizations (NGOs) often echo these dismal assessments of pervasive corruption in China. In 2005, Transparency International deemed China one of the most corrupt places on earth, ranking it 78th out of 158 countries surveyed and behind such countries as Saudi Arabia and Syria. It is impossible to measure precisely the costs that corruption imposes on China's economy, but each of the few serious estimates available suggests that the burden is considerable. One expert has surmised that by the late 1990s, the "predatory exactions" squeezed out of average citizens by CCP officials totaled more than $12 billion per year.[36] In 2004, the OECD estimated that total economic losses caused by corruption among officials ran as high as 5 percent of GDP ($84.4 billion).[37] Likewise, a recent study by the Carnegie Endowment for International Peace estimates that the direct costs of corruption in China amounted to roughly $86 billion in 2003, or 3 percent of GDP. This means that corrupt officials stole more money from state coffers in 2003 than was spent on education nationwide in 2006.[38] Illegal privatization of state enterprises and other assets probably cause the greatest losses to government revenue.[39]

---

[34] Chinese Academy of Social Sciences, *Blue Book of Chinese Society: 2004*, Beijing: Social Sciences Documentation Publishing House, 2004, pp. 37–45.

[35] Dong Liang, "Relationship Between Officials and the Masses," *Blue Book of Chinese Society: 2004*, Chinese Academy of Social Sciences, Beijing: Social Sciences Documentation Publishing House, 2004, pp. 34–36, especially Table 2.

[36] Andrew Wedeman, "Budgets, Extra-Budgets, and Small Treasuries: Illegal Money and Local Autonomy in China," *Journal of Contemporary China*, Vol. 9, No. 25, 2000, pp. 489–511.

[37] Organisation for Economic Co-operation and Development, *Governance in China: Fighting Corruption in China*, Paris, 2005a.

[38] Minxin Pei, "Corruption Threatens China's Future," Carnegie Policy Brief 55, October 2007, p. 2.

[39] Manion, 2004, pp. 113–114.

Unemployment and Regional Income Disparity. China is in a race to create jobs. Its economy not only must accommodate growth in the labor force generated by natural population increases and rural-to-urban migration, but must also provide jobs for the millions of workers laid off from SOEs and the agricultural sector. In the late 1990s, it was losing that race, as downsizing in the state sector outpaced job creation in the private sector by almost four million jobs per year.[40] In 2005, the official figure for unemployment—which is limited to urban areas and measures only a fraction of those actively seeking work—climbed to a record high of 8.4 million.[41]

Hu Angang, a well-known Chinese economist, has employed Western methods to estimate actual unemployment in China. He concludes that in recent years, China's real urban unemployment has run almost three times as high as the official rate, soaring as high as 11–12 percent. Hu's estimates indicate that job seekers in Chinese cities could number as many as 25 million.[42] It has been estimated that open and disguised unemployment in rural China amounts to about 23 percent of the total labor force.[43] China's workforce is expected to swell to 822 million people by 2010, adding eight million to nine million workers per year to the labor pool. If half of the people currently working in the agricultural and state sectors are laid off or otherwise choose to transition to the private sector during that period, then there will be a need to create 20 million to 30 million jobs per year.[44]

A related source of social unrest and discontent with CCP rule is the ever-widening gap between the newly emergent middle class of the coastal cities and the masses of peasants and migrant workers that live

[40] Thomas G. Rawski, "Recent Developments in China's Labour Economy," November 20, 2003, Table 5.

[41] National Bureau of Statistics of China, *China Statistical Yearbook 2006,* Beijing: China State Statistical Press, 2006.

[42] Bergsten et al., 2006, p. 32.

[43] Charles Wolf, Jr., K. C. Yeh, Benjamin Zycher, Nicholas Eberstadt, and Sungho Lee, *Fault Lines in China's Economic Terrain,* unpublished RAND Corporation research, p. xvi.

[44] Douglas Zhihua Zeng, "China's Employment Challenges and Strategies After the WTO Accession," World Bank Policy Research Working Paper 3522, February 2005, p. 8.

throughout the countryside and in shanty towns around major cities. A 2005 UN report warns that the income gap between the urban and rural areas in China, which is among the highest in the world, threatens to undermine social stability.

China: A High-Tech Economy?

Discussions of China's future tend to focus on how quickly the country will start to climb the value-added ladder and transition from an economy focused on low-cost manufacturing to one that fosters high-tech innovation. The Chinese government has set ambitious goals in this regard and no doubt envisions a future role for China in the global economy that rivals those of the United States, Japan, Germany, and other technologically advanced states today.

In the late 1990s, China began to develop an electronics and information technology (IT) sector by attracting investment from Asian and Western high-tech companies that were searching for a low-cost production locale. China's cost competitiveness and large labor pool have allowed it to become a major center of global IT hardware production. However, most of the IT production facilities in China are foreign-owned and are engaged in labor-intensive final assembly. The value-added stages of production often occur outside of China, and key components and technologies are almost always not indigenously designed. With a few notable exceptions, Chinese IT companies are generally oriented toward the domestic market and are not yet considered major players in the global IT industry. An estimated 90 percent of Chinese IT and electronics exports are produced by foreign-invested enterprises.[45]

The development of a robust high-tech economy in China will face at least two major constraints in the near-term. The primary constraint is that China suffers from significant human-capital issues, including brain drain and low-quality tertiary education. By some estimates, China has the world's largest brain-drain problem. A 2007 study conducted by CASS found that seven out of ten Chinese students who enroll in universities abroad never return to China. Since 1978,

---

[45] Bergsten et al., 2006, p. 105.

1.06 million Chinese students have gone abroad to study and only 275,000 have returned.[46] Although noticeable improvements in Chinese university education and research have been made in recent years, the quality of most Chinese universities lags far behind that found in educational institutions in most Western and middle-income Asian countries. The government has promised to increase education spending, but total current government expenditures on education amount to only 2 percent of GDP.[47] Reports of widespread absenteeism in primary and secondary schools across the nation—a rising trend driven by local budget cuts that force public schools to charge high tuition rates—as well as recent indicators of resurgent rural illiteracy suggest that average education levels in China may actually be on the decline.[48] This decline could prove to be an enormous stumbling block to building an advanced, knowledge-based economy in China.

Further evidence of the difficulty of developing China's high-technology sector comes from a 2005 McKinsey & Company study that reports that the nation may soon face a "talent shortage" so severe that it could "stall not only its economic growth, but also its migration up the value chain."[49] Multinational companies have reportedly begun to complain about how difficult it is to find graduates of Chinese universities with the skills required of jobs in the service occupations. In a nationwide survey of companies that hire Chinese workers in nine different occupations, McKinsey found that "fewer than 10 percent of Chinese job candidates, on average, would be suitable for work in a foreign company." Likewise, McKinsey believes that there are only about 3,000–5,000 Chinese managers nationwide who are able to work comfortably in an international professional environment. For this reason, McKinsey judges it highly unlikely that China will develop a strong

---

[46] "China Fears Brain Drain as Its Overseas Students Stay Away," *The Guardian* (London), June 5, 2007.

[47] Bergsten et al., 2006, p. 50.

[48] Maureen Fan, "Illiteracy Jumps in China, Despite 50-Year Campaign to Eradicate It," *The Washington Post*, April 27, 2007.

[49] Diana Farrell and Andrew J. Grant, "China's Looming Talent Shortage," *The McKinsey Quarterly*, No. 4, 2005, p. 1.

sector for offshore IT and business-process services: China lacks the requisite talent to support these industries.[50]

Much has been made of the fact that China currently produces four times as many university graduates with engineering degrees and 50-percent more engineers at the masters and doctorate level than the United States.[51] This point of comparison is not insignificant, but it is important to remember that China is starting from a much lower base than the United States and that the quality of engineering training in China has come under question by many of the multinationals that hire Chinese engineers. In 2005, McKinsey found that although China has a population of 1.6 million young engineers, "China's pool of young engineers suitable for work in multinationals is just 160,000—no larger than the United Kingdom's. Hence the paradox of shortage amid plenty."[52] The problem is attributed to the failure of the Chinese education system to prepare its engineers for the practical realities of the workplace. The Chinese system tends to emphasize theory at the expense of Western-style problem solving through group projects. According to McKinsey, China's "talent shortage" and its excess pool of 150 million unskilled rural workers (who are only suited for employment by manufacturers) have created conditions in which China remains "decades away from developing a consumer-oriented economy."[53]

The second major constraint that will undercut China's efforts to develop a high-tech economy is the government's failure to provide protection for intellectual property rights (IPR). Despite China's rise as a global trading power, Beijing still tolerates IPR infringement rates that are among the highest in the world. The central government has refused to introduce criminal penalties sufficient to deter further IPR infringement, and the few steps taken to date have been mostly cosmetic and have had no real impact on infringement activity. China

---

[50] Farrell and Grant, 2005, p. 4.

[51] Bergsten et al., 2006, p. 103.

[52] Farrell and Grant, 2005, p. 4.

[53] Farrell and Grant, 2005, p. 4.

has stepped up its funding of R&D activities in recent years, hoping to spark a wave of innovation. In 2003, China's total R&D outlays amounted to 1.1 percent of GDP. In absolute terms, however, China's R&D expenditures constitute only about one-tenth of U.S. spending.[54] The ceiling of return on rising R&D expenditures will remain low as long as the personal financial and professional rewards derived from advancements in product designs and new technologies are not protected by a reliable IPR regime. An environment in which IPR violations are pervasive and tolerated is almost certain to strongly suppress indigenous innovation.

Demographic Trends and Related Problems

Aging and Social Security

Unlike many other factors that guide China's development trajectory, demographic change is one variable whose effects we can predict with a reasonable degree of certainty. Within the next two decades, China will be saddled with an unenviable distinction: It will become the first major developing country to experience rapid aging before it becomes a moderately developed economy. In other words, China will get old *before* it gets rich.

A population age structure favorable to development has facilitated China's recent economic boom. Since 1990, China has reaped the benefits of a youthful population in which both young and old dependents represent a relatively small share of the overall population. In 2000, for example, over 70 percent of the Chinese population was between the ages of 15 and 64—nearly 10 percentage points higher than the average in middle-income countries.[55] With 1960s baby boomers powering the workforce, China will continue to enjoy a favorable dependency rate through 2015, at which time the retirement of the oldest baby boomers will begin to usher in the inevitable consequences of years of falling fertility rates and rising longevity.

---

[54] Naughton, 2007, pp. 352–353.

[55] Naughton, 2007, pp. 172–173.

According to projections by the UN, the working-age population in China will begin to shrink after 2015.[56] In some parts of the country, most notably the southern province of Guangdong, reports of labor shortages have already begun to surface.[57] In fact, a recent study by the Guangzhou Municipal Statistics Bureau predicts that the working population (people ages 15 to 64) in the provincial capital will cease to grow by 2013, that dependents will exceed workers by 2024, and that the workforce will represent less than half of the city's total population by 2034.[58]

This striking pattern of demographic change is expected to emerge across all of China between 2015 and 2030. Western demographers estimate that the number of Chinese over the age of 60 will increase from 128 million in 2000 to 350 million in 2030. This will cause the senior dependency rate to soar to 25 percent. The senior dependency rate is expected to climb every year thereafter until at least 2050 (see Figure 4.6).[59] To appreciate the scale of this problem, consider that the number of elderly Chinese in 2030 will likely exceed the current population of the entire United States by 50 million. Demographers at the UN further confirm this trend. They argue that the elderly could constitute as much as 28 percent of the overall Chinese population by 2040. The ratio of working-age Chinese to elderly Chinese now stands at a healthy 6:1, but according to UN forecasts, that ratio will plummet to a mere 2:1 by 2040.[60] Qiao Xiaochun, a Chinese academic who specializes in population studies, has projected an age distribution of China's population through 2050 that is somewhat more optimis-

---

[56] United Nations Population Division, "World Population Prospects: The 2006 Revision Population Database," last updated September 20, 2007.

[57] Jim Yardley and David Barboza, "Help Wanted: China Finds Itself with a Labor Shortage," *New York Times*, April 3, 2005.

[58] "Guangzhou Threatened by Labor Shortage," *China Daily*, September 4, 2007.

[59] U.S. Census Bureau, Population Division, *International Data Base*, last updated June 16, 2008.

[60] United Nations Population Division, "World Population Prospects: The 2004 Revision Population Database," 2004.

Figure 4.6
Ratio of Working-Age Chinese (ages 15–59) to Elderly Chinese (age 60 or over)

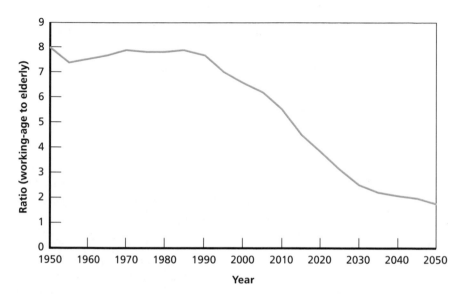

SOURCE: United Nations Population Division, "World Population Prospects: The 2004 Revision Population Database," 2004.

RAND *MG729-4.6*

tic than UN forecasts. Qiao anticipates that the elderly will represent 20 percent of the total population in 2040.[61]

This erosion in the ratio of working-age citizens to retirees has two primary causes. First, the CCP introduced a set of draconian birth-control measures in 1980—often referred to outside of China as the "One-Child Policy"—that have suppressed birthrates for well over two decades and are still enforced. The total fertility rate in China has dropped from 6.1 births per woman in 1949 to only 1.8 births per woman in 2002.[62] A minimum of 2.1 births is generally required for replacement to occur. The second cause of China's demographic

---

[61] Cited in Robert Stowe England, *Aging China: The Demographic Challenge to China's Economic Prospects*, Westport, Conn.: Praeger, 2005, p. 17.

[62] People's Republic of China National Population and Family Planning Commission, home page, copyright 2001.

dilemma is far more positive: rising longevity fostered by strong economic development. Public hygiene, nutrition levels, and health care have improved as a result of wealth creation, and these improvements have contributed directly to a remarkable rise in life expectancy from the low 40s in 1949 to 71.7 years in 2002. By 2025, life expectancy is expected to reach 74.3 years.[63] These changes are making common a 4-2-1 family structure in China consisting of four grandparents, two children, and one grandchild. As older generations move into retirement age, they will create severe economic pressure on the younger generations, who will have to support a number of older dependents.

The financial burden that an aging society will place on only children born under the One-Child Policy is made heavier by China's failure to establish a well-funded social security system in the wake of the privatization of SOEs. The current social security system is too limited in scope to properly address the needs of a rapidly graying society. In 1997, China established a nominally national pension system for urban areas. This system consists of three tiers. The first tier provides a modest, flat-rate pension of 20 percent of average urban wages (regardless of an individual worker's lifetime earnings). In the second tier, employers and employees jointly contribute 10 percent of an employee's salary to an individual retirement account. Participation at these levels is technically mandatory, but enforcement across provinces and municipalities is uneven and the proportion of covered workers in urban areas remains rather low. In fact, only about 16 percent of the economically active population nationwide enjoys coverage.[64] The third tier, by far the most underutilized, consists of enterprise annuity plans that are voluntary, employer-sponsored supplementary pensions.

Rural China lacks a pay-as-you-go social security system for its more than 300 million agricultural workers. Nonagricultural workers in rural areas—a group that in 2001 included 131 million people who work for township and village enterprises, 12 million people who work in rural private enterprises, and the 26 million self-employed—are also

---

[63] World Bank, *World Development Reports, Country Tables: China*, undated.

[64] Bergsten et al., 2006, p. 28.

denied coverage under the current social security system.[65] Millions of urban workers outside of the formal sector of the economy, mostly migrants, are not included in the urban pension system. If the Chinese government fully funds and implements the 1997 national pension plan in urban areas alone, RAND economists estimate that the cost could amount to 3 percent of current GDP and 5.5 percent of GDP by 2025.[66] These estimates do not account for the burdensome additional cost that Beijing would incur if it took steps to provide pension coverage to the multitudes of rural workers who subsist on far lower average incomes than their urban counterparts.

In addition to these enormous coverage gaps, the urban social security system is underfunded. The central government estimates the social security shortfall to be between $122 billion and $244 billion. The Bank of China International published a report that placed the pension shortfall at $850 billion. Credible estimates from the private sector say that the shortfall exceeds $1 trillion.[67] Yan Wang, an economist at the World Bank, argues that the current pension system is unsustainable. He maintains that if the system does not change, payroll taxes will necessarily increase steadily to bring the pension system into balance. According to Wang's calculations, absent any further reform, tax rates would have to rise from 24 percent in 2001 to 27 percent in 2005, then to 45 percent in 2030, and finally to nearly 60 percent in 2050 to keep the social security system solvent.[68] In any case, China's graying society will clearly present unprecedented fiscal challenges to the nation's still-developing economy. Social unrest will undoubtedly spread if Beijing fails to fashion an adequate response to popular demands for a social security net that can support its rapidly aging population.

---

[65] England, p. 78.

[66] Crane et al., unpublished RAND Corporation research, p. 213.

[67] England, 2005, pp. 78, 88–89.

[68] Yan Wang, Dianqing Xu, Zhi Wang, and Fan Zhai, "Implicit Pension Debt, Transition Cost, Options, and the Impact of China's Pension Reform: A Computable General Equilibrium Analysis," The World Bank, Policy Research Working Paper, February 2001.

All of this suggests the likelihood of slowed economic growth and greater pressure on China's central government to provide health care and retirement incomes to the elderly.

Gender Imbalance

Coupled with a long-standing cultural preference for male children, the One-Child Policy has led to another demographic dilemma that could contribute to social unrest in the near future. China currently suffers from a gender imbalance that is unprecedented in countries that have not fought a major war in nearly 30 years. The global norm for the male-to-female sex ratio at birth is roughly 105:100. In 2000, the most recent year for which national census data are available in China, the male-to-female sex ratio for the infant-to-four-year-old age group was reportedly 120.8:100.0. Compared with global averages, this suggests that more than 12 million girls were unaccounted for by the 2000 census; most females were likely aborted upon discovery of the sex of the fetus.[69]

As a result of this gender imbalance, the Chinese government estimates that by 2020, there could be as many as 30 million men of marriageable age who will be unable to find a spouse.[70] Such a situation could fuel social problems, including petty crime, prostitution, human trafficking, drug abuse, and HIV/STD transmission. Some political scientists even argue that large numbers of "surplus males" could create social conditions in which it is advantageous for the Chinese government to build a massive military force that provides an outlet for the young men's aggressive energies.[71]

Floating Population and Urbanization

Migration from the rural areas of China to the bustling coastal cities has proceeded at an astounding pace over the last decade: The population of urban dwellers has swelled by an estimated 200 million. The

---

[69] Census data cited in Naughton, 2007, pp. 171–172.

[70] James Reynolds, "Wifeless Future for China's Men," BBC News, February 12, 2007.

[71] Valerie M. Hudson and Andrea M. den Boer, *Bare Branches: Security Implications of Asia's Surplus Male Population*, Cambridge, Mass.: MIT Press, 2004.

strain that this massive movement of people has placed on city infrastructure, such as water and sewage systems and public transportation, has been at times almost crippling.[72] Mass migration has also caused demand for urban housing to soar. Internal mobility has thus increased exponentially despite the continued existence of a household registration system [*hukou*] that was in part designed to prevent unauthorized migration within China. The lure of higher wages in the cities has given rise to a migrant "floating population" [*liudong renkou*] that consisted of approximately 145 million people in 2000—a number equal to roughly half the entire U.S. population at the time.[73]

Despite this internal movement, roughly 60 percent of China's population still resides in rural areas. Although the rural economy accounts for 40 percent of China's workforce, it produces only 15 percent of the nation's total economic output.[74] As China's economy continues to develop, further migration from rural areas and small towns to larger cities is inevitable. Continued growth in the urban population will compel the government to invest more in infrastructure to sustain economic growth in these municipalities.

Large-scale migration in China also has potential political ramifications. As the population moves, so will ideas and opinions. Since most migrants are poor and undereducated, their migration opens the door for the dissemination of political views and information within a segment of the population that is not likely to have access to the Internet or other means of gathering information about distant places. As migrants return home periodically, there is potential for further dissemination of new information to the countryside.

---

[72] China's cities are completely unprepared for this population influx. For example, fewer than half of Chinese cities provide municipal sewage treatment.

[73] Zai Liang and Zhongdong Ma, "China's Floating Population: New Evidence from the 2000 Census," *Population and Development Review*, Vol. 30, No. 3, September 2004, pp. 483–484. In this monograph, the floating population is defined as individuals who have resided at the place of destination for at least 6 months without local household registration status.

[74] Richard McGregor, "China Must Cut Farming Population, Says OECD," *Financial Times* (London), November 14, 2005.

Public Health

The rapid aging of Chinese society and the massive migration of workers from rural areas to crowded cities are demographic nightmares when viewed in the context of China's broken health care system. Most of China's population has extremely limited access to health insurance because government-sponsored universal health care was abandoned following the economic reforms of the 1980s.[75] The central government has progressively cut its contribution to health care spending, leaving individuals and their families to make up the deficit. The central government's expenditures on health care amounted to less than 1 percent of GDP in 2004.[76] Private spending on health care now represents almost twice as much as public spending as a percentage of GDP. Nicholas Lardy, a renowned economist and China specialist, estimates that only one-seventh of the Chinese population has basic health insurance.[77] In 2003, the central government introduced a Rural Cooperative Medical Scheme intended to provide subsidies for treatment of critical illnesses to more rural citizens. The efficacy of this program is not yet clear, but initial reports indicate that limited funding has severely hampered progress.[78] Official estimates of insurance coverage also reveal an abundance of uninsured Chinese. The Development Research Center of the PRC State Council published a report in 2005 that shows tremendous coverage gaps: Less than half of urban residents and only 10 percent of rural inhabitants are currently covered by the medical insurance system.[79]

Exacerbating the problem posed by such enormous gaps in medical insurance coverage is the skyrocketing cost of medical care in China. For most Chinese, health care is increasingly unaffordable,

---

[75] Yuanli Liu, "Development of the Rural Health Insurance System in China," Health Policy and Planning, Vol. 19, No. 3, May 2004, p. 159.

[76] National Bureau of Statistics of China, 2005.

[77] Bergsten et al., 2006, pp. 28, 51.

[78] Zi Li, "Medical Reform at the Crossroads," Beijing Review, Vol. 48, No. 38, September 2005, p. 23.

[79] Rongxia Li, "The Medical Reform Controversy," Beijing Review, Vol. 48, No. 38, September 2005, p. 24.

and reports suggest that the prohibitive expense is deterring many from seeking medical treatment. Statistics from the PRC Ministry of Health (MOH) reveal that average per capita clinic and hospitalization expenses climbed annually by 13 percent and 11 percent, respectively, from 1997 to 2005; this increase far outpaced growth in per capita income. For example, the municipal government of Zhengzhou, a city of over seven million people and the capital of the province of Henan, estimates that per capita annual medical expenses leaped 244-fold between 1984 and 2004.[80] In a recent nationwide survey, the MOH found that 48.9 percent of respondents report that they typically do not visit a doctor in case of illness; 29.6 percent are not hospitalized when it is medically necessary because the cost is too high.[81] Even childhood immunizations have declined in recent years because patients must finance the shots themselves.[82] Furthermore, prohibitive medical costs prevented many people who may have had severe acute respiratory syndrome (SARS) from reporting to the hospital for treatment during the 2003 crisis.[83] There is also evidence that the growing burden of health care costs being placed on patients and their families has further aggravated poverty in the countryside.[84]

As the 2003 SARS scandal made readily apparent, China's health care system was not prepared to deal with a large-scale epidemic. Although much of the SARS debacle stemmed from political efforts to control public information and maintain domestic stability,[85] the sudden outbreak of illness exposed larger shortcomings in disease monitoring and health care in China. The SARS experience suggests that a similar failure of the health care system to control a future epidemic could have significant political or economic effects, or both. For

---

[80] R. Li, 2005, p. 23.

[81] R. Li, 2005, p. 21.

[82] "China's Health Care: Where Are the Patients?" *The Economist*, August 19, 2004.

[83] Peter Wonacott, "Ailing Patient: In Rural China, Health Care Grows Expensive, Elusive," *Wall Street Journal*, May 19, 2003.

[84] See Liu, 2004, p. 159.

[85] See Joseph Fewsmith, "China and the Politics of SARS," *Current History*, Vol. 101, No. 656, September 2003, pp. 250–255.

instance, Beijing has previously tried to conceal the severity of epidemics such as HIV/AIDS and SARS by misreporting the number of those infected. Both epidemics nonetheless prompted significant unrest. HIV-positive villagers, many infected through blood donation, protested for greater government assistance.[86] Villagers panicking during the SARS outbreak sometimes rioted, attacking officials and medical facilities during 2003.[87] Participation in some of these SARS-related riots numbered in the thousands.[88] The scale of the SARS riots raised concern in the central government, which was forced to take steps to respond to criticism that it mishandled SARS. These measures included firing local government and health officials and promising to increase the transparency of disease reporting.

Even though fewer than 6,000 cases of SARS were ultimately identified, the 2003 SARS crisis cost China an estimated 0.2 percent in decreased GDP growth.[89] This decrease proves that health and disease have significant implications for China's economic growth. A severe epidemic could decrease the labor supply, alter the retiree-to-worker dependency ratio, and choke population growth. Life-shortening diseases could undermine investment in human capital. Serious epidemics could also affect investment and business in the Chinese economy by decreasing the local savings rate or by scaring away foreign investors.[90]

Even if China avoids another major epidemic, health care is poised to become an enormous resource sponge that will absorb an increasingly large portion of government expenditures from now until at least mid-century. Starting in 2015, the world's most populous country will

---

[86] "China Detains Four HIV-Positive People Asking for Help," Agence France-Presse, July 15, 2004; and "Group Says China Increased Arrests, Violence Against HIV Positive Protestors," Agence France-Presse, July 9, 2003.

[87] Bill Savadove, "SARS-Wary Villagers Riot, Attack Officials," *South China Morning Post*, May 6, 2003.

[88] Jonathan Watts, "Punch Line: SARS Sparks Chinese Riots," *The Guardian* (London), May 6, 2003.

[89] "SARS Impact Serious but Not Overwhelming: APEC Report," *Xinhua News Agency*, October 17, 2003.

[90] Wolf et al., 2003, p. 67.

turn gray at a brisk pace. Since data from advanced market economies suggest that average health care costs rise rapidly with age, we expect that pressure on China's medical system is likely to grow as the numbers of the elderly increase. Chinese leadership will be hard-pressed to ignore demands for medical insurance from elderly citizens when their numbers balloon to between 20 percent and 30 percent of the population. Moreover, a purportedly socialist government that governs a nation whose tradition of revering the elderly is proud and deeply rooted can hardly hope to survive if it fails to convince its citizenry that every effort is being made to meet the population's basic health care needs.

Environmental Degradation

China's environmental problems are immense in their magnitude and severity. The ongoing Chinese environmental crisis both erodes economic growth and poses legitimacy problems for a central government unable to protect its population from environmental harm. A recent World Bank study carried out in conjunction with the PRC State Environmental Protection Agency (SEPA) found that the "health costs of air and water pollution in China amount to about 4.3 percent of its GDP." The non–health related effects of pollution were judged to cost an additional 1.5 percent of annual GDP. This yields a total of 5.8 percent of GDP (approximately $100 billion) that China loses every year to air and water pollution.[91] Direct damage to GDP caused by pollution is caused by acid rain damage to crops, medical bills, decreased productivity, disaster relief spent on floods, and the implied costs of resource depletion. China's environmental woes are in no way isolated by region or limited to only certain types of pollution. The overall quality of the air, water, and land in most of China falls well below— often dangerously below—international standards for environmental health and safety.

---

[91] World Bank, *Cost of Pollution in China: Economic Estimates of Physical Damages*, Washington, D.C., February 2007.

Beyond the direct economic costs, widespread pollution and water scarcity reduce the quality of life in China and undermine confidence in the government's ability to solve problems. China's environmental troubles reportedly forced the politically difficult relocation of an estimated 20 million to 30 million farmers in the 1990s; by 2025, 30 million to 40 million more may have suffered the same fate.[92] In a 2005 interview, a senior Chinese environmental official bluntly criticized the country's wasteful development strategy:

> We are using too many raw materials to sustain this growth. To produce goods worth $10,000, for example, we need seven times more resources than Japan, nearly six times more than the United States, and perhaps most embarrassing, nearly three times more than India. Things can't, nor should they be allowed, to go on like that.[93]

Water Scarcity and Pollution

The growing scarcity of water probably constitutes the most under-appreciated threat to sustainable economic growth and social stability in China. The quality of water is also swiftly deteriorating due to widespread pollution; this quality problem in turn intensifies issues stemming from scarcity. Water shortages are most severe in China's dry northern regions, where rapid depletion of groundwater, excessive diversion of rivers, and natural aridity have created chronic scarcity. As of 2003, China's north had 14 percent of the nation's total supply, yet the region is home to 538 million people, roughly 42 percent of China's total population.[94] As a result, inhabitants of this region experience severe water scarcity, receiving only 757 cubic meters of water per

---

[92] Elizabeth Economy, *The River Runs Black: The Environmental Challenge to China's Future*, Ithaca, N.Y.: Cornell University Press, 2004, p. 82.

[93] Andreas Lorenz, "The Chinese Miracle Will End Soon: Spiegel Interview with China's Deputy Minister of the Environment," Patrick Kessler, trans., *Der Spiegel*, No. 10, March 7, 2005.

[94] Zmarak Shalizi, "Addressing China's Growing Water Shortages and Associated Social and Environmental Consequences," World Bank Policy Research Working Paper 3895, April 2006, p. 7.

person per year. This level of water availability is well below the international definition of water scarcity—1,000 cubic meters of water per person per year—used by the UN and World Bank. In fact, it is only one-tenth of the world average. Boasting abundant rainfall and a water availability rating of 3,208 cubic meters of water per person in 2003, China's southern regions are far more blessed with water resources than their northern neighbors.[95] Nevertheless, as with most of the country, industrial pollution has ensured that not even the southern regions are exempt from water issues.[96]

Nationally, China's water resources are expected to reach the level of "water stress" (2,000 or fewer cubic meters per person per year) by 2010, assuming the population continues to grow at the 2003 rate. The national figure for water availability in 2003 (2,180 cubic meters per person) shows that China already stands on the brink of the official stress level. National levels of water availability in China are less than one-third of the average in developing countries and are just one-fourth of the world average. It is estimated that more than 400 of China's 600 cities regularly lack sufficient water, and roughly 100 are confronted by a serious water shortage. In 2000, Chinese scientists calculated total water shortages for the nation at 38.8 billion cubic meters;[97] they also predicted that the water shortage could exceed 50 billion cubic meters a year (more than 10 percent of current annual consumption) by 2020.[98]

Even more dangerous is the fact that all of China's major river systems suffer from high levels of industrial and agricultural pollution. A 2003 SEPA survey found that roughly one-quarter of the water from the Pearl and Yangtze Rivers was deemed unfit for human or industrial consumption. Yet, these two rivers fared the best by far. More

---

[95] Shalizi, 2006, p. 7.

[96] In 1993, the stretch of the Yangtze River near Chongqing was found to have dangerously high levels of chromium, mercury, lead, ammonia nitrogen, petroleum, acidity, coliform, and oxygen depletion. Many environmentalists believe that the Three Gorges Dam will trap pollutants around the densely populated Chongqing area. See Naughton, 2007, p. 493.

[97] Shalizi, 2006, pp. 9–10.

[98] Economy, 2004, p. 71.

than 70 percent of the water in five of China's seven major rivers—the Yellow, Hai, Liao, Huai, and Songhua rivers in the north—could not be designated for any beneficial use. This figure was even larger (over 80 percent) for the Hai and Huai rivers.[99]

A 2007 survey by the Nanjing Institute of Geography and Limnology suggests that the situation on the Yangtze, China's longest river, is deteriorating and that the damage could be irreversible. Field research conducted by the scientists from Nanjing found that more than 370 miles of the Yangtze were in "critical condition" and that nearly 30 percent of the river's tributaries were "seriously polluted."[100] The Yangtze absorbs nearly half of the country's total wastewater, nearly 80 percent of which goes untreated.[101] It is estimated that more than 75 percent of the water flowing through the urbanized stretches of China's rivers is unsuitable for drinking or fishing.[102] A 2007 study by the OECD concludes that one-third of the total length of China's rivers are "highly polluted." The same is true of 75 percent of China's lakes and 25 percent of its coastal waters.[103]

In the three major lakes targeted by a late-1990s cleanup campaign—the Tai, the Hu, and the Dianchi—a 2003 SEPA study discovered that more than 70 percent of the Tai Lake monitoring stations and 100 percent of the Hu and Dianchi Lakes monitoring stations indicated that water was at best suitable only for irrigation purposes and that much of the water was unsuitable for any use.[104] Conditions

---

[99] Economy, 2004, p. 69; and Shalizi, 2006, p. 11. Economy describes in great detail the alarming situation in the Huai River basin. In 1994, a surge of polluted water killed scores of fish, caused widespread illness, and forced businesses along the river to shut down. Beijing spent $7.2 billion on a cleanup campaign and vowed it would be done by 2000. Local governments, however, protected factories from expensive upgrades and shutdowns, and the Huai River today remains one of the most polluted in the world.

[100] "China: Yangtze Is Irreversibly Polluted," Associated Press, April 15, 2007.

[101] Jiangtao Shi, "Pollution Makes Yangtze 'Cancerous,'" *South China Morning Post*, May 31, 2006.

[102] Economy, 2004, p. 18.

[103] John Vidal, "Dust, Waste, and Dirty Water: The Deadly Price of China's Miracle," *The Guardian* (London), July 18, 2007.

[104] Economy, 2004, p. 69.

in Lake Tai have likely worsened since the SEPA survey: Water supplies for four million people were cut off for over a week in June 2007 due to an algae bloom that engulfed 70 percent of the lake's surface.[105]

China's water-pollution problems can be traced to a number of sources. The World Bank identifies the following primary culprits: industrial and municipal wastewater discharges; agricultural runoff from low-quality chemical fertilizers, pesticides, and animal manure; and the leaching of solid waste.[106] The government had made progress in recent years in regulating industrial wastewater from large factories, but the overall volume of industrial and municipal wastewater is still on the rise. From 2002 to 2003, these two kinds of wastewater discharge rose a combined 4.7 percent.[107]

Smaller factories, including the township and village enterprises (TVEs) that dominate much of China's rural landscape, still lack even basic facilities to treat wastewater. TVEs could be responsible for as much as half of all industrial wastewater in China. The same rural communities that typically host TVEs also suffer from growing amounts of agricultural wastewater produced by the relatively recent popularization of low-quality chemical fertilizers and pesticides.

Municipal wastewater remains a significant problem in most major cities. In 2003, the central government reported that only 42 percent of municipal wastewater was treated before being dumped into local rivers.[108] Outside of the major cities, an estimated 200 million people live in towns that possess no sanitation system other than "pipes that lead wastewater to the nearest ditch."[109] The degradation of water quality brought about by wastewater pollution exacerbates problems with water scarcity downstream in the river systems that it affects. It also complicates efforts to recycle water in areas suffering from scarcity.

---

[105] Peter Ford, "Tai Lake Algae Bloom Cuts Off Water to Millions in China's Jiangsu Province," *Christian Science Monitor*, June 4, 2007.

[106] World Bank, *Clear Waters, Blue Skies: China's Environment in the New Century*, Washington, D.C., 1997, p. 90.

[107] Economy, 2004, p. 70.

[108] Cited in Naughton, 2007, p. 491.

[109] Economy, 2004, p. 71.

About 60 million people in China already have limited access to water for their daily needs, and at least 600 million people consume water that is contaminated with animal and human waste. Moreover, only six of the 27 largest cities in China supply drinking water that complies with SEPA standards.[110]

Pollution has exerted pressure on already-stressed water supplies, but demand from both urban residential and industrial sources has also risen sharply since economic reforms were implemented in the early 1980s. Urban residential demand for water has more than quadrupled from 7 billion cubic meters in 1980 to 32 billion cubic meters in 2002. Over the course of the same 22-year period, industrial water demand rose by an astounding 250 percent from 46 billion cubic meters to 114 billion cubic meters.[111] Beyond the normal effects of economic growth on water demand, runaway water usage is also attributable to government-controlled pricing that sets water service and wastewater disposal costs at artificially low levels.[112] These low prices eliminate economic incentives to minimize water use or control pollutants.

Rising demand and polluted rivers have led to excessive depletion and unsustainable exploitation of most sources of groundwater in northern China. Chinese scientists estimate that half of the 60 billion cubic meters of total nonrenewable groundwater in the North China Plain was depleted during the second half of the 20th century. The remaining supply could be exhausted in some areas within as little as 15 years.[113]

Desertification

Desertification has increased rapidly in China since the 1970s, posing an enormous problem for a country already struggling with water scarcity. In 1994, one Chinese scientist calculated the direct economic

---

[110] Economy, 2004, pp. 68–69.

[111] Shalizi, 2006, pp. 9–10.

[112] Jun Ma, *China's Water Crisis [Zhongguo Shui Weiji]*, Norwalk, Conn.: Eastbridge, 2004, p. 28.

[113] Ma, 2004.

losses resulting from desertification at $6.53 billion per year.[114] This sum has undoubtedly increased in recent years as desertification has spread. Desertification is largely anthropogenic in nature. Although natural factors have contributed somewhat to the problem, the primary causes of desertification in northern China are, by percentage, overcultivation (25 percent), overgrazing (28 percent), deforestation (32 percent), and vegetation destruction caused by industrial construction and misuse of water resources (9 percent). The principal natural factor causing desertification, sand-dune encroachment, accounts for just 6 percent of the total problem and is more easily reversible than the anthropogenic causes.[115]

More than one-quarter of the land mass controlled by the PRC is now desert.[116] Large-scale surveys of desert areas conducted in 1994 and 1999 by the PRC State and Forestry Administration reveal that desert land expanded by 52,000 square kilometers (approximately 20,000 square miles, the size of Mississippi) in that five-year period.[117] Sandstorms, the most prominent result of this trend, have recently increased in frequency and severity throughout northern China. Sandstorms and dust storms often dramatically reduce visibility for days at a time in Beijing, forcing school and airport closings and elevating airborne particulate matter to life-threatening levels.

In addition to causing disruptive sandstorms, desertification is also threatening vital ecosystems and contributing to the degradation of China's arable land. Only 15 percent of the PRC is arable, and virtually all of this land is exploited. Thus, further reduction caused by desertification must be viewed seriously. After all, the CCP has historically placed great priority on maintaining self-sufficiency in the provision of food. Recent reports indicate that China's arable land decreased

---

[114] Tao Wang and Wei Wu, "Sandy Desertification in Northern China," in Kristen A. Day, *China's Environment and the Challenge of Sustainable Development*, Armonk, N.Y.: M.E. Sharpe, 2005, p. 239.

[115] Wang and Wei, 2005, pp. 236–237.

[116] Economy, 2004, p. 66.

[117] Naughton, 2007, p. 498.

by 760,000 acres in just the first ten months of 2006.[118] If this trend toward desertification is not arrested, further loss of arable land will occur, thereby threatening the livelihoods of the people who depend on arable land for survival. Desertification will also diminish sources of surface water and groundwater, increase poverty, and result in more disruptive sandstorms.

Air Pollution

Air quality in most Chinese cities has been severely compromised by the prominence of heavy industry in China's economic development strategy and by the country's strong reliance on coal for energy production. The situation is likely to further deteriorate as coal consumption increases and personal automobiles become more popular. A 2002 study by the World Bank found that, of the world's 20 cities with the worst air quality, 16 are located in China.[119] The central government admits that nearly two-thirds of its cities do not meet national air-emission standards, which are more lax than those of many developed nations. According to a 2004 study conducted by SEPA, 75 percent of urban residents in 340 cities where air quality is monitored were breathing polluted air.[120]

The primary form of air pollution in China is particulate matter, which has been widely implicated in respiratory and pulmonary diseases. In 2002, SEPA carried out air-quality tests in over 300 cities nationwide and found that nearly two-thirds of those cities had levels of total suspended particulate matter that exceeded the standard set by the World Health Organization for acceptable exposure.[121] China also has the world's highest levels of sulfur dioxide emissions, a chemical responsible for acid rain. China's coal-fired factories spewed 25.5 million tons of sulfur dioxide in 2005, more than double the level deemed safe. As a result, government sources admit that highly destructive acid

---

[118] "Pollution Hits China Farmland," BBC News, April 23, 2007.

[119] Economy, 2004, p. 72.

[120] Bergsten et al., 2006, p. 54.

[121] Economy, 2004, p. 72.

rain now falls across one-third of China's landmass, with some areas seeing 100-percent acid rain.[122] By 2002, an estimated 30 percent of China's arable land was affected by acid rain, and nearly $13.3 billion in annual damages were sustained from acid rain's corrosive effects.[123] Some neighboring nations, such as South Korea and Japan, blame China for acid rainfall in their own countries.

In addition to contributing to the world's highest levels of sulfur dioxide emissions, China's previously discussed addiction to coal has catapulted the nation to the head of the pack in another undesirable ranking: CO_2 emissions. The IEA believes that China will surpass the United States as the world's largest producer of greenhouse gasses within the next two years.[124] The IEA originally expected this development to occur in 2010, but China's rapid emissions growth has compelled the agency to revise its forecast twice in less than a year.

China's determination to push ahead with a coal-intensive energy strategy ensures that its greenhouse gas emissions will continue to grow exponentially. China's CO_2 emissions are projected to increase fivefold by 2030, with power-generation plants accounting for approximately 55 percent of total emissions. By 2030, the IEA believes that China's CO_2 emissions will be 41 percent greater than those of the United States; it also projects that China's increase in emissions between 2000 and 2030 will nearly equal the emissions growth seen in the rest of the industrialized world.[125]

According to a July 2007 report in *The Financial Times*, the CCP has tried to hide the health consequences of such severe air pollution from the Chinese people. The World Bank reportedly removed references to the excessive number of deaths caused by air and water pollution in China from a forthcoming report following repeated protests from SEPA and the MOH. China already has the world's highest rate of chronic respiratory disease, but the World Bank report also found

---

[122]"Third of China 'Hit by Acid Rain,'" BBC News, August 27, 2006.

[123]Economy, 2004, p. 72.

[124]Barbara Lewis, "China About to Become Biggest CO_2 Emitter: IEA," Reuters, April 18, 2007.

[125]Bryan Walsh, "The Impact of Asia's Giants," *Time Magazine*, March 26, 2006.

that approximately 700,000 Chinese die prematurely each year due to the effects of air pollution; another 60,000 perish annually from water pollution.[126] Beijing feared that the dissemination of such "sensitive" information would incite social unrest. Despite major consequences for public health, the central government plans to build 562 new coal-fired power plants by 2012—an expansion equal to nearly half the world's total number of existing coal-fired power plants.[127]

Equally worrisome are the early signs that many Chinese who live in more-developed cities are beginning to embrace ownership of personal automobiles as a status symbol. Major cities such as Beijing, Shanghai, and Guangzhou already exhibit regular patterns of severe traffic gridlock. Cars designed and manufactured by Chinese companies typically emit 10–20 times more carbon monoxide, nitrogen oxide, and hydrocarbons than U.S. and Japanese models of comparable size.[128] At present, per capita vehicle ownership in China is remarkably low, and the total fleet nationwide is estimated at 24 million vehicles. With only 22 cars per 1,000 people, China is well below the global average of 114 vehicles per 1,000 people and dramatically below the U.S. level of 764 cars per 1,000 people. However, if recent trends persist, the number of vehicles on the road in China could total 100 million by 2020.[129]

If China achieves a per capita vehicle ownership rate equal to that of the United States today, the country's roads will be clogged with 600 million to 800 million cars in 2050—a number equal to the entire current world fleet.[130] Barring a revolution in clean fuel technology, it is difficult to imagine how China's environment could sustain such high levels of car ownership. The greenhouse emissions produced by such a

---

[126]Richard McGregor, "750,000 a Year Killed by Chinese Pollution," *Financial Times* (London), July 2, 2007.

[127]Bergsten et al., 2006, p. 54.

[128]Economy, 2004, pp. 74–75.

[129]Bergsten et al., 2006, p. 54.

[130]Economy, 2004, p. 74.

vast number of vehicles would pose catastrophic risks to the health of the planet.

Enforcement

Awareness of China's environmental plight is increasing among central government officials and within various segments of the general population. Beijing has ratified 16 international environmental treaties, but absent the embarrassment of a news-making crisis, enforcement of both its international commitments and domestic regulations is exceedingly weak. Enforcement is rare because responsibility for enforcement has been delegated to local officials who often prioritize economic growth over environmental stewardship.[131] SEPA's perceived lack of political clout and its paucity of personnel are other important factors inhibiting the enforcement of environmental regulations. Last but not least, the ruling party's general distrust of international and Chinese NGOs and its refusal to allow a free press prevent the emergence of two effective whistleblower communities that could otherwise monitor local officials' compliance with Beijing's environmental mandates.

As previously mentioned, China's reliance on coal comes at a high price. By some estimates, China will become the world's largest single emitter of CO_2 in 2007. China's decision to push ahead with a coal-intensive energy strategy ensures that this situation will only worsen. China's CO_2 emissions are projected to increase by 500 percent by 2030; power-generation plants will account for approximately 55 percent of total emissions.

Supply Trends

Since becoming a net importer of oil in 1993, China has strongly diversified its import sources. China now draws its 3.6 million barrels per day of crude imports from almost every major oil-producing region. Angola emerged as China's largest and most important oil supplier in

---

[131] For a detailed discussion of the politics of China's "environmental responsibility system," see Elizabeth Economy, "Environmental Enforcement in China," in Kristen A. Day, ed., *China's Environment and the Challenge of Sustainable Development*, Armonk, N.Y.: M.E. Sharpe, 2005, pp. 102–120.

2006 (followed by Russia and Saudi Arabia).[132] Although Venezuela sells a relatively small share of its total production to China, Caracas has announced plans to significantly expand oil exports to China by the end of the decade and is thus likely to emerge as a new major source of China's oil supply.

China's efforts to secure its energy supply by buying stakes in foreign assets appear to be of limited success. Presently, only two of the exploration and development projects in which Chinese oil companies have invested generate more than 100,000 barrels of oil per day for the Chinese home market. These investments are a greater than 90-percent stake in Aktobemunaigaz (in Kazakhstan) and a majority stake in Sudan's Greater Nile Petroleum Operating Company. In 2005, the total contribution of Chinese-owned assets abroad to China's oil imports amounted to less than 8.5 percent.[133]

That said, however, China's demonstrated willingness to bypass global oil markets and seek exclusive drilling rights to obtain some semblance of supply security does not bode well for future energy cooperation with the United States, especially in times of crisis. If the two nations do not adjust their energy consumption patterns, it is likely that competition for access to affordable energy sources could prove to be a point of friction in future relations.

Conclusion

China is living out a Faustian bargain. With its growth-at-any-cost policy, it has been the beneficiary of wealth and military might. It has enjoyed three decades of robust economic growth that has secured it a prominent place in the global economy. China's once largely peasant population is becoming increasingly urbanized. The PLA is transforming itself from a huge defensive ground force with antiquated equipment to a modern military capable of limited power projection.

---

[132] Energy Information Administration, 2006a.

[133] Eurasia Group 2006, "China's Overseas Investment in Oil and Gas Production," prepared for the U.S.-China Security and Economic Review Commission, October 16, 2006.

However, in exchange for this wealth and might, China has sacrificed its environment, its natural resources, its energy future, and the well-being of its elderly and rural residents. The production of coal, China's main source of energy, could peak in the next two decades and is expected by some analysts to drop significantly below current levels by 2030–2040. Deferred costs and the vestiges of a command economy could begin to create a drag on further economic development if the central government fails to take reform to its logical end. The educational and judicial systems needed to foster high-tech innovation and support advanced service industries have not yet materialized. In the next two decades, China's baby boom generation will begin to retire. Many of these baby boomers, like China's elderly today, will lack retirement incomes and will therefore compound China's economic problems. Public access to health care is increasingly scarce and unaffordable. Unhealthy air and undrinkable water are common. Cropland is turning into desert, and water sources are being rapidly depleted. China's heavily populated northern regions, including Beijing, already suffer from severe water scarcity; by 2020, the water shortage nationwide could exceed 50 billion cubic meters a year. Some of these conditions are irreversible; all are worsening and intertwined.

The systematic political reform that is needed to effectively address these problems is unlikely to occur in the near future because the CCP has proven vigilant and tireless in its efforts to forestall meaningful political reform. Although it is clear that the CCP will attempt to remain in power by stunting the growth of civil society, the party will face increasingly assertive demands from migrant workers, rural residents, and others left behind by a corrupt and unfair system. In contrast, relatively strong support for maintaining the political status quo is likely to persist among the urban middle class as long as the economy retains its strength. This unlikely source of support for the CCP stems from Beijing's deft adoption of policies aimed at convincing the urban business elite that the CCP is the best protector of its economic interests. In 2001, Jiang Zemin began this courtship when he endorsed the formal admission of entrepreneurs into the CCP. In the years that followed, the CCP eased the tax burden on domestic enterprises and entrepreneurs and showered public servants with frequent, substantial salary increases and heavily subsidized housing in comfortable, newly built apartments.[134]

The CCP's successful employment of sophisticated policy measures to co-opt the emerging middle class has undermined the notion that China's economic opening to the world would fuel domestic political change. Large-scale, organized opposition to CCP rule is unlikely as long as economic growth continues and urban Chinese enjoy a rising standard of living. In the event of an economic crisis, however, the urban middle class could easily turn on the CCP if presented with a viable political alternative. It is not entirely clear whence such an alternative would emerge, but most plausible scenarios involve new leadership coming out of a split within the party ranks.

It is impossible to determine when exactly an economic downturn might become a politically destabilizing crisis, but the breaking point could occur well before China falls into a full-blown recession. Since the CCP's legitimacy rests almost entirely on the expectation that the party will deliver robust economic expansion, it is conceivable that even a short period of anemic growth could trigger disruptive political unrest and widespread opposition to single-party rule. In the event that its legitimacy is called into question, the CCP will probably stoke the flames of nationalism in a last-ditch effort to justify continued single-party rule, even if this tactic risks conflict with its neighbors or the United States. Nationalism is a potentially explosive wildcard in China's future that could dramatically shape its international behavior, especially if Chinese public opinion blames external factors or foreign actors for China's internal insecurities. A vivid illustration of this danger emerged in the months leading up to the 2008 Beijing Olympics, as protests and boycotts directed at French supermarket Carrefour erupted across all of China in response to protests in Paris against China's brutal occupation of Tibet. The central government in Beijing and the state-controlled media played up reports of "anti-China" prejudice in the West and encouraged nationalists to avail themselves of the Chinese Internet (heavily censored under normal conditions) to

---

[134]Jonathan Unger, "China's Conservative Middle Class," *Far Eastern Economic Review*, April 2006, p. 28.

vent their anger and to organize "patriotic" protests.[135] For decades, the CCP has nursed historical grievances among the populace and directly facilitated the rise of a victim mentality that blames Western and Japanese "imperialism" for China's weaknesses.[136] The widespread sense that China is presently on course to reclaim its rightful place as the center of the civilized world has taken on a self-perpetuating momentum that nearly spun out of control during anti-Japanese riots in the spring of 2006. The United States should be prepared to manage the misunderstandings and miscalculations that will likely result from the nationalist posturing of an economically turbulent China.

In the immediate future, in order to maintain power, the CCP will likely continue pressing ahead in a gradualist manner that attempts to balance internal pressures and sustain economic growth without undertaking deeper reforms that would end the lingering distortions of a still partially command-driven economy. Further reform of SOEs and the financial sector (the largest sources of these distortions) is unlikely to materialize given the loss of control over critical sectors of the economy it would entail and the ideological crossing of the Rubicon that it would represent. To paraphrase Thomas Jefferson, the CCP seems to have concluded that continuing to hold on to the ears of a wolf is better than letting go.

Although defenders of the status quo clearly constitute the majority in the halls of power in Beijing, other factions differ in their assessments of how the country should proceed in the coming years.[137] From China's reformist "right" comes pressure to minimize the state's active intervention in the economy and, to a lesser extent, to adopt political reforms that increase accountability and representation within the party. From China's reactionary "left" comes pressure to maintain China's socialist heritage by introducing measures to protect millions of

---

[135] Jill Drew, "Protests in China Target French Stores, Embassy," *The Washington Post*, April 20, 2008, p. A23.

[136] For a more in-depth treatment of this subject, see Peter Hays Gries, *China's New Nationalism: Pride, Politics, and Diplomacy*, Los Angeles: University of California Press, 2004.

[137] See Edward Cody, "Hu Set for Second Term at China's Helm," *The Washington Post*, October 14, 2007.

farmers and laid-off workers from an extraordinarily corrupt and predatory market economy. This latter choice risks slower economic growth and "backwardness" while preserving CCP control over the economy.

In the longer term, China may be forced to change course. In turning to the right, the CCP might attempt to secure its future by focusing internally on domestic issues to reduce internal instabilities. The set of reforms most likely to emerge from such a scenario include the provision of a national retirement-pension system and universal health care for the elderly. It is possible that a turn to the reformist right could also result in efforts to reduce the number of money-losing SOEs, which not only drain resources from enterprises that are potentially more efficient and profitable but also obstruct reform of the dysfunctional financial sector. If China chooses to focus on these costly economic and social reforms along with securing its energy supplies, attention to these domestic priorities would most likely create a drag on further military modernization. This course of action is more likely to create opportunities for cooperation with the United States, but increased economic competition for energy resources in the GME could undermine relations if it proves to be too fierce.

In turning to the reactionary left, the CCP might attempt to manage growing internal instabilities by squashing dissent and lowering the public's expectations for further reforms. The CCP might maintain or strengthen its control over the economy, thus perpetuating market distortions and slowing GDP growth. A turn to the reactionary left would increase China's opportunities to further enhance its military power, use that power to secure energy supplies and raw materials, and resist the United States militarily. In the extreme, a Lebensraum policy might be adopted as a means of acquiring fresh land and resources and rallying the Chinese populace around the flag. Turning to the left could thus lead to increased confrontation between China and the United States. However, slower economic growth associated with the Chinese left's economic agenda could lessen competition for energy resources in the GME.

It is too early to judge with confidence whether China is more apt to become a threat, a competitor, or an ally. To be sure, China's aggressive military modernization has transformed and strengthened a traditionally ground force–centric military by giving far greater emphasis to naval and air forces. A military long postured to fight extended wars of attrition is now focusing on defending Chinese interests off its littorals in short, high-intensity campaigns. On one hand, China is displaying new, potentially threatening military capabilities, including the ability to destroy satellites in low-earth orbit, that raise questions about its long-term intentions. Ongoing cyber-attacks against the United States are another indicator of this ominous trend. Likewise, China's diplomatic and economic support for Iran, the Sudan, Burma, and other rogue regimes is also viewed with displeasure by the international community. Perhaps most worrisome is how China's promotion of a model of authoritarian politics blended with market reform is gaining currency in the developing world as an alternative to the "Washington Consensus." On the other hand, the witch's brew of domestic challenges that are likely to beset China in the coming two decades could discourage further military development. Strong economic ties between China and the United States will also likely mitigate against sources of greater competition or conflict.

In the next 10–15 years, China must confront the complex and intertwined problems of dwindling energy resources, increasing energy needs, slower economic growth, the emerging unmet needs of its elderly, severe pollution, and shortages of vital natural resources (such as water). Inertia caused by 30 years of stellar economic growth could very well tempt Chinese leaders to believe that they can "cheat the devil" by finding new routes to growth that diminish the environmental and demographic consequences of their Faustian bargain. Evidence suggests, however, that China's current growth strategy is unsustainable, since the myriad problems cited above will only worsen. Continuing the policies of the present is a precarious proposition that could ultimately undermine the stability that the CCP's gradualist approach has sought so desperately to maintain.

CHAPTER FIVE

China's Near Abroad

This chapter examines the status of China's current relations with its neighbors. First we examine current and near-term relationships and their effect on China. Then we explore how these relationships might change in the medium to far term. China's evolving relations with its neighbors will, of course, significantly influence China's strategic situation and resource expenditures. For purposes of this discussion, we divide China's near abroad into three regions: the east and northeast, the southern tier, and the west.

The Current Situation

The PRC enjoys good or very good relations with its southern tier of neighbors (i.e., Vietnam, Laos, Thailand, Burma, the Philippines, and India). Whereas in earlier decades there was considerable tension or even open warfare between the PRC and most of those nations, today China has mutually beneficial and growing economic ties with the southern tier. China's economic-political model, which combines many of the benefits of capitalism with single-party authoritarian rule, has considerable appeal in Vietnam, which is adopting a similar approach. The Philippines and Thailand, who in the past regarded the PRC with considerable suspicion and even fear, today see China as the main economic engine in Asia. They increasingly regard China as a primary trading partner and a means to enhance their own economic growth. Even India, who fought China in the 1960s, now regards the PRC as an increasingly important trading partner. Trade between China and India has been increasing rapidly in the past five years, reaching $25 billion in 2006.[1] At present, China and its southern tier of neighbors are enjoying the mutual benefits of increased economic activity.

Although conflict along China's southern region in the near term appears unlikely, there is the prospect of future tension between China and India as they struggle for influence along the Indian Ocean littoral. Until recently, India was the dominant power in the region. Now, China is making major strategic investments in the key regional states of Burma, Bangladesh, Sri Lanka, Tibet, and Pakistan, who are India's neighbors. These Chinese investments include expanding military and economic ties. China is now making a major investment to improve the region's transportation infrastructures, with particular emphasis on Burma, Tibet, and Pakistan.

This investment in South Asia is consistent with China's domestic economic strategy of "going West" as it expands its access to energy supplies in the GME. China's external investments—in the new Gwadar port in Pakistan, for example, and in a major pipeline that will run through Tajikistan, Uzbekistan, and Turkmenistan toward Iran—are coupled with a major investment to improve China's domestic transportation infrastructure in the western part of the country. Numerous road and rail projects in western China are intended to facilitate the inflow of oil and natural gas from the "stans" and Iran and should promote greater economic activity in the historically poor and underdeveloped western part of the country.

Whereas the strategic situation in the west and south is today quite favorable for China, the east and northeast remain areas of tension and potential danger. The fate of Taiwan remains a key question. Although the PRC has generally been very patient over the Taiwan issue, hoping for an eventual peaceful reintegration of the island with the mainland, there remains the real possibility that some miscalculation on the part of one or more of the interested parties could lead to a sudden crisis. The current situation is one of political standoff and

---

[1] "Reaching for a Renaissance: A Special Report on China and Its Region," *The Economist*, March 31–April 6, 2007, p. 10.

waiting while the economic and financial integration between China and Taiwan moves apace.

The current government of Taiwan, the Kuomintang, is attempting to reach some degree of political accommodation with the PRC. This process will clearly take years, and may end if the present Taiwanese leadership is removed from office in the next series of elections.

Although the Taiwan problem is an important and difficult one for China, the situation in North Korea (also known as the Democratic People's Republic of Korea [DPRK]) is potentially even more challenging, if only because it is more complex. In spite of his dependence on China for economic and political support, DPRK leader Kim Jong Il has elected to remain defiant over the issue of nuclear weapons. The DPRK's acquisition of a nuclear capability has, in turn, pushed the Japanese toward greater military expenditures and a more assertive political-military mindset. This partial "awakening" of Japan in areas beyond economic activity is not a welcome development in China.

Through recent Six-Party Talks, Chinese diplomacy was able to convince the DPRK to begin the process of denuclearizing its nuclear arsenal and nuclear-weapon industry. After a delay that lasted much of 2006 and 2007, the DPRK has once again shut down its operational nuclear reactor at Yongbyon and allowed International Atomic Energy Agency (IAEA) inspectors to return to confirm the reactor's closure. Simultaneously, $25 million in frozen North Korean bank assets were returned to the DPRK while South Korea (also known as the Republic of Korea [ROK]) began to supply the North with a massive quantity of fuel oil. These are the opening moves of a denuclearization agreement reached in February 2007. It is widely expected that the timeline associated with Pyongyang's agreement to give up its small nuclear arsenal and account for the dismantlement of its entire nuclear-weapon production process in a verifiable manner will be protracted.

It is important to note that the China of today, an authoritarian capitalist state and rising great power, is not ideologically aggressive in the mode of the Soviet Union. China is far more integrated into the global economy than the Soviet Union ever was. At present, China's main goal vis-à-vis its neighbors is to minimize the possibility of conflict while increasing its economic opportunities. The CCP's ability to maintain single-party political dominance over the domestic scene depends on its ability to provide the Chinese people with growing prosperity and lasting stability. Today, China needs and wants to be part of the global economic system to support the strategy of frenzied growth required to allow the country to absorb the several hundred million Chinese who are migrating from rural areas into major cities. Thus far, that participation has been hugely to the PRC's advantage, as evidenced by China's productive relations with its near abroad.

Future Trends and Possibilities in the West

China's considerable attempts to develop infrastructure in its western region and its growing involvement in the "stans" (including Pakistan) and Iran will continue to increase at least through the medium term. China's energy needs are increasing rapidly, and despite plans to significantly increase use of domestic coal and nuclear power, the PRC will have to import ever-greater quantities of oil and natural gas in the future. As China's dependence on imported oil and natural gas continues to increase, the PRC's economic and political interests in the Middle East will grow. It is highly likely that China will increase its strategic ties with Iran in all dimensions while also sustaining a robust economic relationship with the rest of the region.[2]

Because a significant portion of the imported oil and gas will be traveling overland into western China, the Chinese can reasonably be expected to gradually increase their security presence in the "stans" and possibly as far west as Iran. This greater level of military involvement could range from selling weapons to various nations in the region, to supplying Chinese military advisors, to stationing Chinese military forces temporarily in some of the countries. This deeper engagement is justified by the fact that the region is now of vital strategic-economic interest to the PRC. Aside from generating anxiety on the part of Washington and the EU, this increased military involvement will cause Moscow to view Beijing as a major player in the region. To a great

---

[2] "Reaching for a Renaissance," 2007.

extent, what is going on today and in the foreseeable future is a 21st-century version of the 19th-century "Great Game" in Central Asia. On one hand, China and Russia may effectively collaborate through the Shanghai Cooperation Organization (SCO) to limit U.S. and EU influence. On the other hand, Russia will be very alert to the prospect of China displacing Russian strategic and economic influence in what Moscow deems its own near abroad.

Future Trends and Possibilities in the South

The most important medium- to far-term variable along China's southern tier is China's evolving relationship with India. As mentioned above, the Indians and Chinese are currently increasing trade with one another, and there are little or no near-term indications that serious hostility between the two countries will arise. There is, however, no guarantee that this situation will continue.

Several factors could lead to a worsening of Sino-Indian relations. These include (1) growing Chinese involvement in Burma, which the Indians are watching with considerable interest, and (2) China's degree of commitment to Pakistan, India's bitter rival. Since the 1950s, the PRC has supported Pakistan over India. If the tension between India and Pakistan worsens (over Kashmir, for example), and if China elects to provide support to Islamabad, India's attitude toward the PRC would almost certainly harden. If there is simultaneous Sino-Indian competition for influence in Burma, a general downturn in the relationship of the two countries could follow.

As mentioned earlier, China's dependence on energy imported from the Middle East in the medium term will be considerably greater than it is today. Although some of the imported oil and gas will be moved overland into western China via pipeline, a considerable portion will still travel by ship across the Indian Ocean to Chinese ports (and possibly Yongon, Burma, if the Chinese succeed in developing the infrastructure there to allow oil to flow northward into southern China). If the Chinese begin to view the Indians as unfriendly, the Chinese could feel compelled to develop military capabilities in the

Indian Ocean region (possibly by developing agreements with Burma to protect their energy sources and transit routes). At a minimum, the emergence of a Chinese naval presence in the Indian Ocean littoral will likely stimulate a major Indian naval buildup. As China expands its economic and military ties with India's neighbors, India will probably respond by greatly expanding its ties with key Southeast Asian states (such as Thailand, Malaysia, Singapore, Indonesia, and Vietnam). An important issue for the future is the extent to which India will seek to expand its political and military ties with Australia, Japan, and, especially, the United States, all of whom will see New Delhi as a strategic counterweight to Beijing's growth as a great power in East Asia.[3]

Future Trends and Possibilities in the East and Northeast

Taiwan

The three basic potential outcomes of the Taiwan situation in the medium to long term are summarized below.

A protracted (decades-long) continuation of the current status quo between the PRC and Taiwan. In this scenario, varying periods of cooperation and tension occur between the two states, but the situation does not reach the level of either integration or war. This scenario posits continued, gradual financial and economic integration of the two states.

A commonwealth arrangement similar to the "one nation, two systems" arrangement that was reached with Hong Kong. In this scenario, Beijing feels that it has essentially reintegrated Taiwan into the PRC; the leaders of Taiwan, meanwhile, claim with considerable justification that they have "fought off" attempts by the mainland to subjugate the people of Taiwan. Such an arrangement could occur before 2020, but is more likely to evolve over a longer period of time. As in the previous scenario, Taiwan and the mainland gradually become more integrated both financially and economically, although

---

[3]   Jane's Information Group, "Going Soft, China's Alternative Route to Regional Influence," *Jane's Intelligence Review*, July 1, 2007.

in this case the integration would be more formal. A Hong Kong–like arrangement in Taiwan would have to include even greater freedom for the island than is the case with Hong Kong.

Warfare between the PRC and Taiwan. The leadership of the PRC has, on several occasions, made it clear that any Taiwanese effort to declare independence would be considered a *casus belli*. This third scenario would, of course, be highly destabilizing, producing major economic consequences in Asia and possibly throughout the world (particularly if the United States elected to support Taiwan with military force). Both China and the United States continue to modernize their military establishments in the Western Pacific with this contingency, however remote, in mind.

The third scenario, war between the mainland and Taiwan, could take one of several forms: The Chinese could attempt an all-out invasion of the island; an economic blockade with submarines and aircraft; or a series of coercive air, naval, and missile attacks in an attempt to force the Taiwanese to acquiesce to Chinese demands. The nature of such a conflict would, of course, have major implications for what capabilities the U.S. Navy would need. Unfortunately for U.S. planners, the Chinese could pursue any one of the options listed above.

The three scenarios have very different implications for the PRC, including very major implications for Chinese military spending. If the Chinese believe that an eventual accommodation can be reached with Taiwan without resort to force, China can reduce its level of military expenditure (at least in terms of a Taiwan crisis). On the other hand, if the leadership of the PRC believes that the use of force might be necessary—and that that use includes the possibility of having to defeat a U.S. attempt to intervene—the resource implications for China are very different.

North Korea

In northeast Asia, the main issue is the DPRK. Several possibilities could play out over the medium term.

DPRK survival. In this scenario, with Chinese and South Korean economic assistance, the DPRK remains an impoverished but nuclear-armed nation. This possibility rests on the assumptions that the DPRK

(1) fails to give up its operational nuclear arsenal and (2) does not suffer any subsequent severe economic penalties. This scenario would result in continuing tension in the region, with the possibility of war remaining. Japan might reconsider its decision not to acquire a nuclear arsenal.

A DPRK-ROK confederation. In this case, some form of political agreement unites the two Koreas. This scenario is likely to flow from a successful denuclearization of the DPRK followed by a dramatic increase in economic aid from South Korea. In this scenario, the DPRK becomes a key transportation node in Northeast Asia. From a Chinese perspective, this scenario has both advantages and disadvantages. The chance of a war on the Korean peninsula would be dramatically reduced, to Chinese satisfaction. However, the Chinese would have to deal with an emerging united Korea on its doorstep. Beijing would undoubtedly try to ensure that the U.S. military presence on the Korean peninsula faded away after reunification. Tokyo would respond by trying to improve its political and economic ties with the peninsula to ensure that the emerging united Korea did not become a Chinese satellite. This scenario would probably further accelerate the strengthening of U.S.-Japanese security ties.

A DPRK collapse. North Korea is an impoverished nation that is straining to maintain its huge military. The nation already receives considerable food aid from various nations, and has experienced serious famine in the past. In this third scenario, the DPRK "implodes," possibly due to a severe economic downturn or during a period of political transition after Kim Jong-Il's departure from the scene. In this scenario, the ROK intervenes to prevent complete chaos in the North, and eventually forces reunification. The Chinese would clearly be alarmed by this situation, especially since they would probably be the recipients of a large number of refugees coming from the DPRK. Concern on the part of both China and the ROK over the security of the DPRK's nuclear arsenal (assuming the denuclearization plan fails and the North still retains atomic weapons at the time of its collapse) could result in tension and possible conflict between the two countries. Beijing would probably demand that a united Korea be free of U.S. military presence. The Japanese would be very concerned about such dramatic events on the Korean peninsula.

A second Korean War. In this final scenario, a second Korean War emerges as the DPRK reneges on the February 2007 denuclearization deal and commits provocative acts of active nuclear proliferation (such as the sale of weapons-grade fissile material to Iran or to a nonstate actor such as al Qaeda). These actions lead to an international response with harsh economic sanctions that receive some Chinese and South Korean support. The unpredictable North Korean regime then provokes a crisis that results in war. A war on the Korean peninsula would be a disaster for the region, since nuclear-weapons use would become a distinct possibility.[4]

Japan

The final key variable for China in the north and northeast region is the issue of "whither Japan?" There remains to this day considerable bitterness in China over Japan's brutal occupation of the 1930s and 1940s, which caused the deaths of millions of Chinese. Although the two countries have become trading partners due to mutual economic interests, there remains a strong undercurrent of distrust and even hatred. It is not clear how Sino-Japanese relations will evolve in the medium to long term. There are several major possibilities.

Japan as Brazil. In this case, the Japan of the future concludes that China is the rising power in Asia and that American presence and influence are declining. The Japanese reach an understanding with the PRC similar to the de facto relationship between the United States and Brazil. In the latter case, Brazil, a country with enormous potential power, has realized that it can take certain liberties in the Western Hemisphere but that it should not overtly challenge U.S. power. In this scenario, Japan reaches a similar conclusion regarding its relationship with a rising China. The plausibility of this scenario depends on the progress of denuclearization on the Korean peninsula and on evidence that the U.S. extended deterrent commitment to Japan has faded.

Japan as the UK. In this scenario, Japan concludes that its long-term interests are best served by becoming increasingly close to the United States. Japanese leaders decide that China's rise poses a major

---

4 "China: The Shifting Strategy on Korea," Stratfor, April 23, 2007.

threat to Japan's national interests and therefore conclude that Japan should develop even closer ties with the Americans. These ties would parallel the lines of the "special relationship" between the United States and the UK. These closer ties might be part of a larger Asian political-military alliance designed to constrain China's rising power, an alliance that could include India, Australia, and other Southeast Asian nations. The past two Japanese governments have been moving slowly but steadily in the direction of this scenario. A clear example of this trend is the Japanese government's decision to upgrade the Japanese Defense Agency to a ministry. Note that this change in status occurred without provoking any meaningful political opposition in what was in earlier years a distinctly pacifist Japan.

Japan as Gaullist France. In this final scenario, the Japanese conclude that in the midst of growing Sino-U.S. tension and competition in Asia, they need to reduce their reliance on Washington to guarantee their security, but without bowing to Beijing. Therefore, in a manner similar to France in the 1960s, Japan would seek to increase its military, including the very real possibility of developing an independent nuclear deterrent. The collapse of the Korean peninsula denuclearization plan coupled with evidence that the United States remains preoccupied with the GME would provide a powerful incentive for Japan to move in such a direction. As noted above, this national security policy might emerge as part of a Japanese effort to build closer political-military ties with India, Australia, and key Southeast Asian states to act as a counterweight to China's rising strength.

The direction that Japan elects to take in the medium to long term will be of great significance to China. Japan remains one of the four most important economies in the world, along with the United States, China, and the EU. Japan certainly has the economic means to become a significant rival of China in Asia. The consequences for Chinese strategic planning are considerable regardless of which course the Japanese take.

CHAPTER SIX

Strategic Trends in Iran

Iran is a complex country with profound social, political, and economic issues that will continue through 2030. Although Iran's nuclear intentions and President Mahmoud Ahmadinejad's (who has ruled since 2005)[1] actions dominate current policy- and defense-related discussions, the longer-term themes that will affect Iran include its economic structure, population, environmental pressures, and future political direction. Iran is unlikely to pose a significant direct strategic challenge to the United States despite current worries. Domestic pressures will continue to dominate, direct, and often dictate Iran's defensive position. Iran will likely choose to continue its preferred soft-power and unconventional approaches to foreign policy at the regional and international levels given its current and future domestic constraints and capabilities.

The Current Situation

Domestically, Iran faces three broad issues that will affect its defense capabilities over the near, medium, and far terms. These include profound *economic structural weaknesses* caused by rentierism, unemployment and underemployment, ongoing demographic pressures, and energy production and consumption patterns; *political uncertainty* characterized by unresolved tensions between theocratic, military,

---

[1] The next presidential election is scheduled for mid-2009.

and reformist elements; and *environmental vulnerabilities.* These factors, either alone or in combination, will dominate Iran's diverse and often factionalized domestic landscape. While two issues—economics and politics—are theoretically amenable to reform, the potential for change is minimal due to the profound disruptive features of economic reform and the presence of entrenched political interests that will resist efforts on both fronts.

Economic Trends

Economic issues are profoundly problematic for the Iranian regime and will remain so through the far term. Historically,

> changes in the political environment have had a major impact on economic growth in Iran. The periods of relative political stability and absence of major external conflicts (1960–1976 and 1989–2002) are clearly associated with high GDP growth, whereas the political turmoil and war period of 1977–88 was associated with negative growth . . . [T]he average annual growth rate during the 1977–88 period was reduced by 6 percentage points.[2]

In the near term, Iran is once again enmeshed in a period of uncertainty that can be expected to have a negative effect on the country's economy.

Several intrinsic structural features of the Iranian economy will significantly affect Iran's ability to grow over the near, medium, and far terms. Critically, there is neither the political will nor the capacity to restructure the country's rentier economy in the foreseeable future. Iran's economic structure will keep the country highly dependent on oil profits, influence government spending patterns, reduce the likelihood of needed declines in domestic energy consumption, and maintain barriers to foreign direct investment (FDI).

---

[2]   Abdelali Jbili, Vitali Kramarenko, and José Bailén, *Islamic Republic of Iran: Managing the Transition to a Market Economy*, Washington, D.C.: International Monetary Fund, 2007, p. 13.

Although the regime has made strides to address the demographic crisis facing the country, underemployment and unemployment will remain significant and will increase over time as the Iranian society ages. Finally, the structural underpinning of Iran's economy—oil—will present interesting challenges to the country over the next several decades. Ahmadinejad's electoral victory was premised on his populist promise to share the country's oil wealth with every household, yet the oil sector remains underdeveloped and vulnerable to international reactions to Iranian leaders' bombast and nuclear efforts. Coupled with significant domestic consumption patterns, this vulnerability significantly erodes many of the advantages of the country's oil resources, advantages that have been amplified by current international demand and high prices.

Rentierism

Rentierism is the percentage of rents in government revenues. Hazem Beblawi's definition of rents is widely accepted:

> [R]ents (1) come from abroad, (2) accrue to the government directly, and (3) "only a few are engaged in the generation of this rent (wealth), the majority being only involved in the distribution or utilization of it." The third point requires some elaboration: it is not only that a few people produce the wealth, but that the wealth is the result of a windfall that is very largely independent of any efforts made by citizens of the rentier state.[3]

Between 1972 and 1999, the average degree of rentierism in Iran was 55 percent; Kuwait's rate was 88 percent, Saudi Arabia's was 80 percent, and Bahrain's was 59 percent.[4] Although Iran's overall level of rentierism is quite modest compared to the levels in neighboring countries, rentierism's political effects are apparent in the proliferation of

---

[3] Michael Herb, "No Representation Without Taxation? Rents, Development and Democracy," paper, Georgia State University, December 3, 2003, p. 4. It is useful to refer to Beblawi's original argument (Hazem Beblawi, "The Rentier State in the Arab World," in Giacomo Luciani, ed., *The Arab State*, Berkeley: University of California Press, 1990, pp. 87–88).

[4] Herb, 2003, p. 32.

literally hundreds of autonomous or semi-autonomous political actors in Iran, including the Islamic Revolutionary Guard Corps (IRGC); *bonyads*;[5] and multiple overlapping government institutions, such as the military, the *Majlis* (legislature), *velayat-e-faqih* (Supreme Leader), and the President. Each of these competing structures vies for control over the country's resources and seeks to prevent an erosion of political influence.

Events in the past decade are illustrative of rentierism's impact on Iran's economic, social, and political structures. In 1997, President Mohammed Khatami (who ruled from 1997 to 2005) was elected largely by Iran's youth (the so-called Third Force[6]) and women on a reformist platform. Khatami promised market liberalization that would require the government to enact constitutional reforms, lift government subsidies on key commodities (including energy), and attract FDI. However, Khatami was largely unsuccessful in achieving even modest market reforms because of political opposition from the country's many factions and popular unwillingness to forgo critical subsidies.

Ahmadinejad's August 2005 election surprised many observers because Ahmadinejad was largely a political unknown whose conservatism seemed to contradict the general pattern of Iranian politics since 1997. As noted earlier, he was elected on a populist platform that appealed to rural and poor voters; however, it would be a mistake to conclude that he enjoyed only a narrow support base.

Ahmadinejad's bombastic foreign-policy remarks, Iran's ongoing nuclear program, and Iranian involvement with the Iraq War have caused the international community to largely isolate the regime at

---

[5] *Bonyads* are a network of parastatal institutions that are semi-autonomous power centers within the Iranian political system. They were formed after the revolution and endowed with billions of dollars taken from nationalized royal and elite assets. *Bonyads* have established strong personal and institutional relationships with the government and are a key actor, if not arena, for the ideological rivalries that characterize Iran's political system. See Suzanne Maloney, "Agents or Obstacles? Parastatal Foundations and Challenges for Iranian Development," in Parvin Alizadeh, ed., *The Economy of Iran: The Dilemmas of an Islamic State*, London: I. B. Tauris, 2001, p. 148.

[6] See Jahangir Amuzegar, "Iran's Crumbling Revolution," *Foreign Affairs*, Vol. 82, No. 1, January/February 2003, p. 44.

the official and operational levels. Since Ahmadinejad took power, the country has had to contend with growing inflation, depressed housing and stock markets, and growing unemployment. These factors have led to a $50 billion capital outflow between 2005 and 2006, according to the *Economist* Intelligence Unit. Figure 6.1 shows our projections of Iranian per capita GDP through 2020.

By all accounts, Ahmadinejad has been a disaster for Iran's economy because of his domestic policies and foreign-policy rhetoric. Yet public domestic disapproval remains muted in the face of aggressive government coercion that targets opponents in the press, universities, and political organizations.

Ironically, Ahmadinejad was successful in May 2007 at raising gasoline prices and initiating a rationing program; successive governments have considered price increases and rationing programs but have not pursued them because of the anticipated negative political fallout.

Figure 6.1
RAND Projections of Iranian per Capita GDP

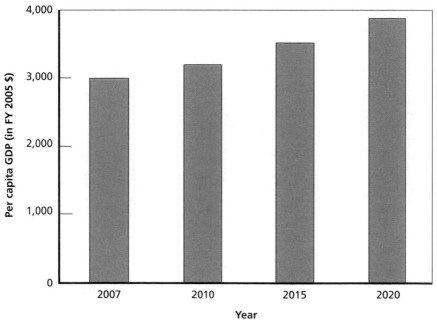

Although the gas-price increase was unpopular and led to limited riot-ing, the government was able to contain the unrest. Assertions that the price increases will undermine Ahmadinejad's 2009 presidential bid are far from widely accepted. We find it interesting that Ahmadinejad felt secure enough to initiate unpopular economic reform. This sug-gests that he may be able to accomplish some economic reforms that his reformist counterparts could not.

Even discounting Iran's current international problems, economic projections do not look encouraging over the far term. Per capita GDP will grow at negligible rates through 2020 (see Figure 6.2). This modest growth will affect military spending and complicate efforts to reform Iran's economy and address the numerous challenges it confronts. As a result, Iran will be incapable of a profound escalation in defense spend-ing and acquisitions. This does not mean that the country's leaders may not institute political changes that will allow the government to redirect the country's resources toward the defense sector. However,

Figure 6.2
RAND Projections of Iranian GDP Growth

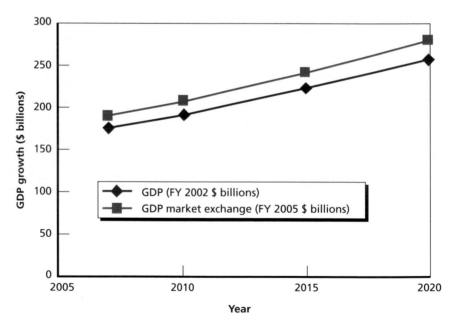

the likelihood of these changes is projected to be low, since popular fallout would be detrimental to Iran's political establishment. Due to its economic structure and projected ongoing economic weakness, Iran will have built-in incentives to maintain its traditional foreign-policy stance, which emphasizes unconventional power and includes support for militias and the use of electoral procedures to gain political influence in key areas (e.g., Hezbollah in Lebanon).

Unemployment and Underemployment

Most analysts concur that the single largest problem affecting Iran's economy is unemployment. Current statistics estimate that this situation will endure through 2011. We believe, however, that this issue will remain a persistent problem for Iran's economy through the far term.

In 2006, Iran's official unemployment rate was 11.6 percent (see Figure 6.3). Unofficial estimates are significantly higher and they reveal important age disparities within Iran's unemployment figures. Notably, the employment picture is weaker than international projections suggest. According to a March 2006 report issued by the Central Bank of Iran, Iran's official unemployment rate was 12.1 percent. Economists believe, however, that the real figure was as high as 20 percent. An Iranian government statement in July 2007 quoted an unemployment rate of 12.4 percent. The *Economist* Intelligence Unit estimates that in 2011, the official unemployment rate will be 14 percent. However, unemployment disproportionately affects those ages 15 to 24. For example, in 2005, the World Bank estimated that unemployment in this age cohort in Iran was 23 percent. Given past trends, Iranian unemployment levels by 2011 among young workers could be as high as 30 percent.

The Iranian government is highly sensitive to unemployment pressures and has sought to create jobs to absorb new labor entrants. However, its capacity to do so successfully has been limited to date. For example, a 2004 estimate notes that 1 million jobs were needed but only 300,000 jobs were created.[7] Similarly, the government sought to

---

[7] Golnaz Esfandiari, "Iran: Coping with the World's Highest Rate of Brain Drain," Radio Free Europe/Radio Liberty, March 8, 2004.

Figure 6.3
Iran's Employment Crisis

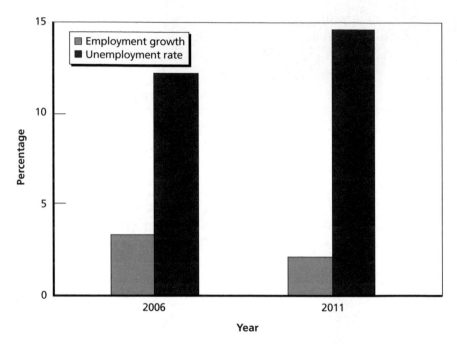

SOURCE: Economist Intelligence Unit, "Country Report Iran," 2006.
RAND *MG729-6.3*

create 3.8 million jobs (760,000 annually) between 2000 and 2005, but only created 2.3 million jobs total. Had the government met its target, Iran's unemployment rate by 2005 would have been 11.5 percent.[8] In July 2007, Iran's government announced plans to reduce unemployment to 8.4 percent by 2010, a goal that requires creating 900,000 jobs annually. Iran is unlikely to reach this target in the mid-term. It is estimated that GDP would need to grow by at least 6 percent per year just to keep unemployment levels constant.[9]

---

[8]   Bill Samii, "Iran: Weak Economy Challenges Populist President," Radio Free Europe/ Radio Liberty, July 21, 2006.

[9]   Abbas Valadkhani, "What Determines Private Investment in Iran?" *International Journal of Social Economics*, Vol. 31, No. 5/6, 2004, p. 457. Other sources suggest a much worse scenario. For example, one report notes that

Complicating Iran's employment situation is the presence of underemployment, defined as workers with high skill levels in low-wage and/or low-skill jobs that do not match their abilities. A related issue in Iran is the need for individuals to have multiple jobs to make ends meet. It is not unheard of in Iran for individuals to hold as many as six jobs. Thus, employment figures do not adequately reflect Iran's economic landscape. Underemployment contributes to the country's brain-drain problem, which has been considered at times to be the worst in the world. For example, in 2004, an estimated 150,000 youth left Iran. As of 2004, an estimated 4 million Iranians lived abroad, with significant communities in North America. Finally, 80 percent of Iran's scientific award winners had chosen to emigrate as of 2004.[10]

Demographic Pressures

The pronatalist policies instituted after the revolution have created a large population cohort that is currently an issue and will continue to present problems through the far term. The country's political and religious leaders initiated an aggressive family-planning program in the 1990s which reduced Iran's population growth rate from an all-time high of 3.2 percent in 1986 to 1.2 percent by 2001,[11] a rate the country sustained through 2007 (see Figure 6.4).[12] Total fertility (the average number of children born to a woman in her lifetime) has dropped from approximately 7.0 in 1986 to 1.71 in 2007, which is below replacement level and an impressive accomplishment.

---

if Iran's GDP growth only averages historical levels of 4.5 percent—with labor productivity growth of 2.4 percent and a gradual increase of women labor participation to 25 percent by 2010 . . . . [T]he average unemployment rate would then be 23 percent.

See Habib Fetini et al., 2003, "Iran: Medium Term Framework for Transition," Social and Economic Development Group, Middle East and North Africa Region, Washington, D.C.: World Bank, April 30, 2003, p. iv.

[10] Esfandiari, 2004.

[11] Janet Larsen, "Iran's Birth Rate Plummeting at Record Pace: Success Provides a Model for Other Developing Countries," Earth Policy Institute, December 28, 2001.

[12] Population Reference Bureau, "Iran," Web page, undated.

Figure 6.4
Iran's Total Fertility Rate

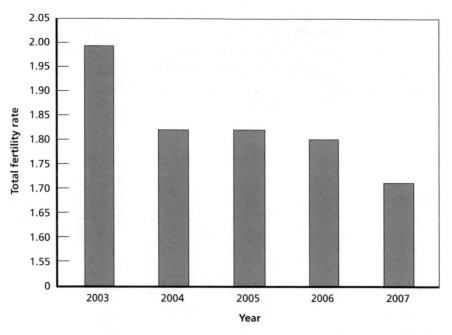

SOURCE: Central Intelligence Agency, "Iran," *World Factbook*, 2003–2007.
RAND *MG729-6.4*

Nevertheless, the long-term costs of Iran's 1980s policy are clear in terms of both the country's youthful and aging population. By 2050, Iran's population over age 65 will have tripled from 2020 levels, while the country's population ages 20 and younger will decline to 20 percent (see Figure 6.5). As a result, Iran will not experience a profound decline in its youngest population cohort through the far term, suggesting that employment challenges in particular will remain fairly constant. More interesting, however, is the graying of Iran's population, which will accelerate significantly between 2020 and 2050. Although this will pose challenges to the Iranian system in terms of social services (such as health care), the figures are roughly balanced by the country's youngest cohorts. Furthermore, there will be a large middle-aged cohort capable of supporting Iran's graying population. Thus, there is unlikely to be a significant change in current pressures related to Iran's population

Figure 6.5
Iranian Population Distribution, 2005–2050

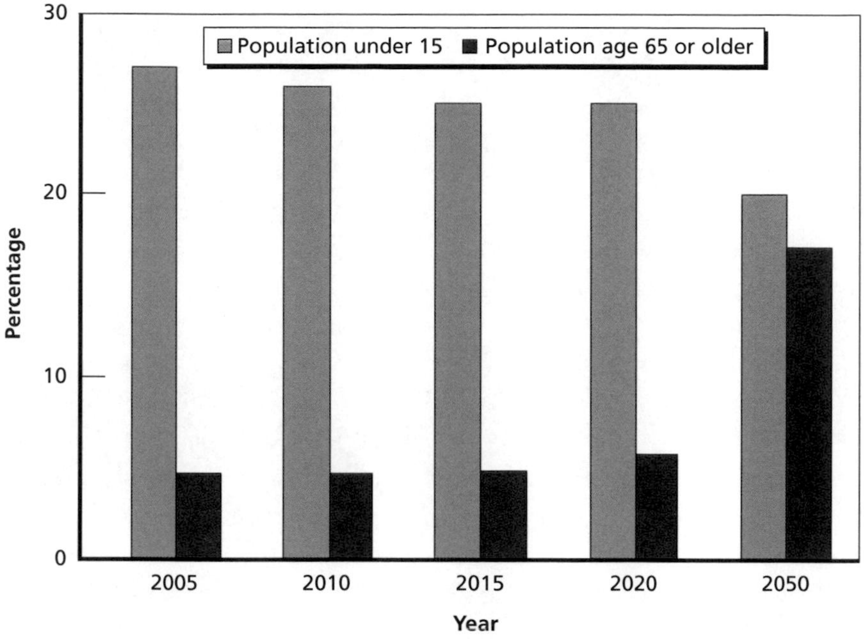

SOURCE: United Nations Population Division, "World Population Prospects: The 2006 Revision Population Database," last updated September 20, 2007.
RAND MG729-6.5

patterns through the far term. Longer-term projections may generate more-favorable outcomes through 2050.

Energy Production and Consumption Patterns

Oil and domestic energy consumption patterns are the primary energy issues facing Iran through the far term. Oil is, and will by all accounts remain, the fundamental pillar of the Iranian economy. Oil revenues account for between 80 percent and 90 percent of total export earnings and between 40 percent and 50 percent of the government's budget.[13] Oil provides government leaders with significant spending latitude,

---

[13] Energy Information Administration, *Country Analysis Briefs: Iran*, August 2006b.

and an important sector of government spending relies on subsidies that oil revenues permit. Two of the most significant subsidy categories in Iran are food and oil. It is estimated that "total Iranian energy subsidies, including gasoline and natural gas, amount to $30 billion, or 15 percent of the country's entire economy."[14]

Oil subsidies keep domestic gasoline costs artificially low and, despite a recent price hike in May 2007, gasoline still costs Iranians only $0.38 per gallon. Subsidies have led to high domestic consumption patterns, lowered profits, and pollution. They also keep free-market forces from improving efficiency in the energy sector in the Iranian economy and stymie opportunities for market reforms and economic liberalization. Experts describe Iran's oil sector as old, inefficient, and underdeveloped. In addition to being riddled with government corruption, the sector is held back by a lack of FDI. This lack of FDI results in more profound psychological effects than economic effects. These factors, in turn, are affected by the regime's muscular political rhetoric and international sanctions. In the near term, Iran's most pressing energy issues concern oil and domestic energy consumption patterns.

One of the most striking features of the Iranian energy sector is that although it is awash in money due to high global oil prices, it is still more cost-effective for the government to import oil rather than rectify the country's refinery problems. A rich subsidy currently prices gasoline at $0.08 per liter ($0.34 per gallon), which has resulted in an 11 percent to 12 percent growth in demand. However, refining gasoline for domestic consumption is unprofitable because of the subsidized price; therefore, imports are favored over refinery expansion. The cost of importing gasoline is small compared with the costs associated with losses due to deferred refinery maintenance. Aggregate refinery leakage accounts for 6 percent of total oil production.[15]

---

[14] "Iran's Decision to Raise Gas Prices Exposes Economic Vulnerability," Associated Press, May 24, 2007.

[15] Roger Stern, "The Iranian Petroleum Crisis and United States National Security," *PNAS*, Vol. 104, No. 1, January 2, 2007, p. 380.

The interrelationships among oil wealth, rentier structural features, domestic consumption, and the country's larger energy profile are visible in other energy sectors, most notably natural gas (see
Figure 6.6). Iranians consume virtually all the natural gas produced
in Iran, thereby reducing the overall amount of oil the government
has to import to meet energy demands (see Figure 6.7). Iran is second
only to the United States as the world's largest gasoline importer (see
Figure 6.8).[16]

Nuclear Energy

One of the most controversial issues for Iran at the international level
centers on its nuclear ambitions. The Iranian nuclear program was officially launched in the 1970s with the help of various Western powers,

Figure 6.6
Iranian Natural Gas

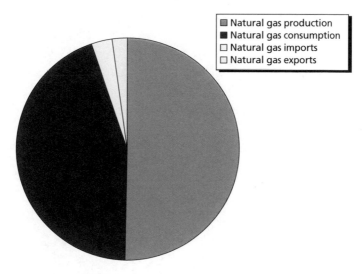

SOURCE: Country Watch, "Iran: 2007 Country Review," 2007.
RAND MG729-6.6

---

[16] Global Investment House, "Iran Economic & Strategic Outlook: Well Diversified Economic Base," June 2007, p. 38.

Figure 6.7
Iranian Electricity

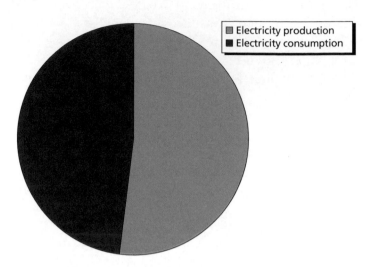

SOURCE: Country Watch, 2007.
RAND *MG729-6.7*

Figure 6.8
Iranian Oil Patterns

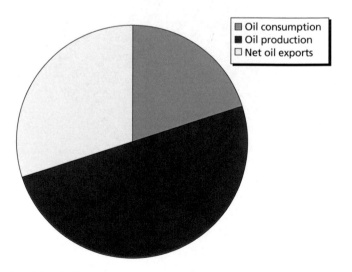

SOURCE: Country Watch, 2007.
RAND *MG729-6.8*

including the United States.[17] Debate surrounds whether the program is designed, as Iran contends, to meet the country's energy needs or whether it is oriented toward meeting the country's security needs.[18] The Iranian government asserts that the program's goal is to develop nuclear power plants, and that it plans to use those plants to generate 6,000 MW of electricity by 2010. Iran's domestic energy consumption is growing annually by 10 percent (versus 3 percent internationally), and household consumption amounts to 56 percent of total consumption in Iran.[19]

Although the former shah publicly proclaimed that the program was for domestic energy generation, former officials within the shah's regime have confirmed that the "program was designed to grant him the option of assembling the bomb should his regional competitors move in that direction."[20] More-current assessments suggest that Iran's nuclear weapons are intended for "deterrence and power projection."[21] The Iranian nuclear program was temporarily suspended after the 1979 Iranian Revolution in part because then-Ayatollah Ruhollah Khomeini saw the indiscriminate effects of nuclear weapons as inconsistent with Islamic canons of war.[22] However, Iran's present-day nuclear program has been developed and sustained by a range of political forces within Iran, including conservatives, pragmatists, and reformists. This cooperation has led to a general public consensus within Iran that is based on the purported linkage between the nuclear program and the country's national interest. Ahmadinejad's current pursuit of

---

[17] Note that a civil agreement was signed between Iran and the United States in 1957 for technical assistance and leasing enriched uranium. See Gawdat Bahgat, "Nuclear Proliferation: The Islamic Republic of Iran," *Iranian Studies,* Vol. 39, No. 3, September 2006, p. 308.

[18] See Bahgat, 2006, pp. 323–324.

[19] "Nuclear Power Plants Will Generate 6,000MW by 2010," *Iran Daily,* November 7, 2005, p. 3.

[20] Colin Dueck and Ray Takeyh, "Iran's Nuclear Challenge," *Political Science Quarterly,* Vol. 122, No. 2, Summer 2007, p. 190.

[21] Dueck and Takeyh, 2007, p. 195.

[22] Dueck and Takeyh, 2007, p. 190.

nuclear capability is consistent with the actions of earlier governments. For example, President Hashemi Rafsanjani reinitiated the country's nuclear program in the early 1990s, and Muhammad Khatami sustained it during his presidency from 1997 to 2005. Currently, Iran's nuclear effort includes research sites, a uranium mine, a nuclear reactor, and uranium-processing facilities (including a uranium-enrichment plant).

There is little doubt that Iran faces significant energy pressures, but the fact remains that Iran's nuclear ambitions have almost wholly moved into the security realm. The security issues surrounding Iran's nuclear program fall outside the intended scope of this analysis, but the negative economic effect of the country's nuclear program is worth noting. Nuclear acquisition has isolated Iran at the international level as a result of sanctions and reluctant foreign investors; these factors will place downward pressure on Iran's economy through at least the near term.

Sociopolitical Trends

Iran is the only successful revolutionary Islamic state. As a result, Islam is the political status quo in Iran, and much of the country's most recent political issues have centered on reform versus conservatism, with the former including both political-religious and economic dimensions.

Following the 1979 Iranian Revolution, Iran's political system was based on a legal order grounded in Islamic law (*sharia*), which was interpreted and sanctioned by the Shi'i *ulama*. The government was structured as a parliamentarian system with a weak executive. Originally, the spiritual leader (*velayat-e-faqih*, hereafter referred to as "Supreme Leader") had supreme authority in the system, but with the death of Ayatollah Khomeini in 1989, the position of the Supreme Leader was given broader executive powers through the revised constitution. Ayatollah Khamenei, who became Supreme Leader in 1989, lacked Khomeini's religious credentials and his tenure has been characterized by a growing effort to consolidate conservative power in Iran, especially since 2004. This process has been largely a reaction to the reformist push that began in 1997 with President Mohammed Khatami's election. Khamenei has allied himself with current and former members of the revolutionary guard and with the volunteer corps (*besiji*), among others, to consolidate conservative power.

One noteworthy change in Iran's political system occurred on December 15, 2006, when, for the first time, both the local-council elections and the Assembly of Experts elections took place on the same day. Local-council elections were institutionalized under Khatami in 1999 to promote bottom-up liberalization. In contrast, the Assembly of Experts is comprised of 86 senior clerics who oversee and support the Supreme Leader and choose his successor. The Assembly of Experts is extremely conservative and was responsible for the suprise appointment of Khamenei (given his anemic religious credentials) as Supreme Leader. The 2006 election "marked a significant change in the electoral process with the advancement of two inherently opposing institutions: one democratic and the other oligarchic." Babak Rahimi provides an interesting argument about the reasons for the electoral change:

> First, at a domestic level, there appears to be an attempt by the conservative authorities, who came to dominate the government since 2004, to give more popular legitimacy to the Assembly of Experts by correlating the re-election of its members with the municipal council elections, as though the two are inherently one and the same. Likewise, since the Assembly of Experts is directly responsible for the selection of the *valyat-e faqih*, the move is made in a way to make Ayatollah Khamenei look as if he is actually an elected member of the government.[23]

In 1997, mostly young voters and female voters elected Khatami president on a reformist platform that was designed to improve economic conditions and loosen some of Iran's religious restrictions. At the time of Khatami's election, democracy was ostensibly a protest vehicle

---

[23] This discussion of the Supreme Leader is from Babak Rahimi, "Iran: The 2006 Elections and the Making of Authoritarian Democracy," *Nebula*, Vol. 4, No. 1, March 2007, pp. 285–290. Also see Ali Gheissari and Vali Nasr, "The Conservative Consolidation in Iran," *Survival*, Vol. 47, No. 2, 2005, pp. 175–190.

that pitted the reformists against the clerics; the reformists enjoyed an important, though not comprehensive, victory. Khatami's government was characterized by modest economic reforms, but true change was prevented by entrenched economic and political interests. Similarly, while some social and political controls were loosened during Khatami's tenure, most notably within the press and university settings, these reforms occurred at a significantly reduced rate during the later period of Khatami's tenure (and ended entirely under Ahmadinejad). The euphoria surrounding Khatami's election and the promise of reform essentially ended when Khatami failed to deliver on his promises; the reputation of reformers was negatively affected at the same time.

The Third Force (Iranians born after 1979) is a common focus of observers interested in political and social change in Iran. There is an assumption that young people are by nature reformist, and this was certainly evident in 1997, when Khatami was elected. However, the Third Force is also a complex political actor that deserves to be better understood. It did not participate in the Islamic Revolution and thus has little recollection of that period. Similarly, the Iran-Iraq War is not meaningful to these young people. They associate Islamic clergy with governmental problems, including corruption, inefficiency, and nepotism. On the other hand, the Third Force is characterized by dualism: Its Iranian nationalism (especially over the nuclear issue) is at odds with its desire for Westernization (especially Western values) and Westernization's associated economic payoffs.

Since 2005, Iran has become less free as the IRGC has increased its influence and domestic and international uncertainties surrounding the nuclear issue and the war in Iraq grow (see Figure 6.9). The freedom of the press has suffered in particular. Ahmadinejad's government has increased public executions of political opponents and has produced a climate of fear by targeting intellectuals, political opponents, and related reformist elements in an aggressive manner since 2006.

Nevertheless, Iran's political institutions are facing significant upward pressures to change. Municipal elections in December 2006 saw major setbacks for conservative candidates aligned with Ahmadinejad, and thus the Iranian political system continues to function in its

Figure 6.9
Iranian Freedom Ranking

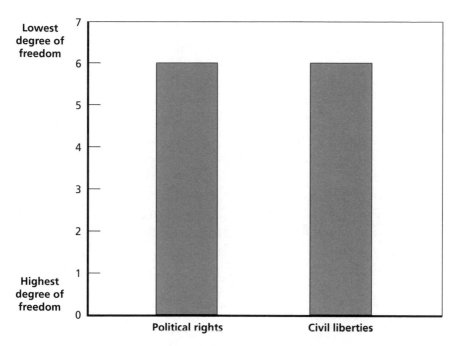

SOURCE: Freedom House, "Iran," *Freedom in the World*, 2006.
RAND *MG729-6.9*

own unusual fashion.[24] Given the multiple competing power structures within the country—which include the judiciary, the military (especially the IRGC), the presidency, the Supreme Leader, the *Majlis*, the *bonyads*, and numerous other actors and agencies—the potential for significant near-term constitutional reform is low. Yet the likelihood that power contestation will continue to be resolved within the country's existing institutions remains high.

---

[24] Stuart Williams, "Ahmadinejad Faces New Setback in Key Poll Battle," *Middle East Times*, December 19, 2006.

Politically, Iran has followed a pattern that oscillates between clientelism and populism;[25] reformism serves as a midway point. Ahmadinejad's 2005 election is consistent with the earlier populist period that followed the 1979 Iranian Revolution; indeed, Ahmadinejad's government's statements emphasize the country's earlier revolutionary, religious, and populist history. Similarly, Khatami's 1997 election captured the reformist middle ground that exists between the clientelist and populist poles. Taking history as a model, the possibility of reformist reassertion during the next presidential election (scheduled for 2009) is low. If Ahmadinejad wins in 2009, a return to clientelism by 2013 is more likely. The pendulum may swing back to reformism after 2020.

In the near term, the most important consideration from a U.S. defense planner's point of view may be changes in Iran's political landscape and their effects on Iran's defense posture. Observers often focus on the contest between the clergy and reformers in Iran's political system, but the more troubling scenario in the near term (and potentially beyond) is the potential battle looming between reformers and the military since the latter seized the upper hand in 2005. Compared with other countries in the region, such as Pakistan and Turkey, Iran has experienced little military meddling in its political affairs. However, this past does not preclude the Iranian military from evolving into a political force. The IRGC in particular has recently gained political

---

[25] Populism is grounded in the 1979 Iranian Revolution and clientelism can be traced to Shiism, rentierism, and the revolution. According to Kazem Alamdari, populism is "a non-class structure with a charismatic leadership. It is a political, ideological, and centralized radical mass movement. Ideologically it represents the declining traditional middle class." Clientelism is the result of three factors: religion (namely Shiism), rentierism, and independent religious organizations (i.e., *bonyads*). According to Alamdari, clientelism is

> a structured relationship between a patron and a client . . . [who] is a subordinate actor who serves his or her patron in exchange for reciprocal rewards. . . . Clientelism is a non-class system with a power structure that consists of separate vertical rival groups rather than horizontal class layers.

See Kazem Alamdari, "The Power Structure of the Islamic Republic of Iran: Transition from Populism to Clientelism, and Militarization of the Government," *Third World Quarterly*, Vol. 26, No. 8, 2005, pp. 1285–1287.

and economic influence in Iran, and the military in general possesses important corporate interests that it will seek to protect and possibly expand.

In the medium to far term, whether the reformers prevail over the military or vice-versa will likely be decided by several actors. The Third Force will be important, but it is neither cohesive nor automatically aligned with democratic or reformist forces. Because male youth must serve in the IRGC, they become indoctrinated and find that their interests are tied to the corporate interests of the military actor. Insofar as the IRGC is a large employer (in part because it exerts influence on businesses), young men are also attached in this way. Elites will continue to be influential informal Iranian policymakers because of their economic resources and ties to the government (e.g., through *bonyads*). Finally, the poor—so often overlooked but so critical to the success of Ahmadinejad's populist platform, which in turn was driven by Khatami's liberal economic and political failures—will continue to be a factor.

Four spoilers will influence all three sectors (the Third Force, elites, and the poor): persistent nationalism, the Iraq war, the lack of institutional reforms, and the fact that bread trumps freedom, economic liberalization steps, and jobs. These factors will influence which of the preceding actors becomes most influential and why, as well as the potential move by Iran toward more market reforms or greater militarization. These factors also shape three potential futures for Iran: continued theocratic rule (i.e., the status quo); further democratization; or militarization. The most likely scenario through the near and medium terms is the status quo. However, through the medium to far term, the likelihood of either democratization (in a more Westernized sense, with significantly reduced clerical influence, reduced rentierism, and market reform) or militarization (with or without clerical influence) increases.

Although the political landscape in Iran changes almost overnight at times, at this juncture what is intriguing and potentially troubling about Iran is that Ahmadinejad has shown a willingness and ability to assert tremendous control over the political system. Some of this power is attributed to Khamenei's continued support, which is viewed as a political effort to balance against Rafsanjani. Nonetheless, it is important to consider that Ahmadinejad enjoys support from other institutions, most notably the IRGC, that has been especially meaningful since 2006. Although it remains to be seen how these political contests will play out over the near and medium terms, it is clear that Iran has taken troubling steps toward greater militarization. Although this trend is worrying, it may hold a silver lining: If Ahmadinejad's government is able to amass enough power and control, it could advance unpopular economic reforms that may have positive implications for Iran through the medium and far terms.

Environmental Trends

Iran faces environmental challenges that are both manmade and natural. Oil production and growing energy consumption patterns have produced, and will continue to produce through at least the medium term, air pollution, environmental degradation, health problems, and growing economic costs associated with environmental degradation. The most pressing environmental problem for Iran is air pollution, and the energy sector is Iran's major source of that pollution (see Figure 6.10).

In 2001, the total health damage from air pollution in Iran was estimated at $7 billion, equal to 8.4 percent of the country's nominal GDP. If reforms do not occur, damage is estimated to grow to 10.9 percent of nominal GDP by 2019.[26] Required reforms involve the elimination of a broad array of energy subsidies, which would result in reduced domestic energy consumption. Barring sweeping reforms, Iran would still benefit from eliminating or reducing fuel subsidies because doing so would reduce consumer demand for gasoline.

Because Iran's socioeconomic environment is dominated by a youthful population and growing urbanization, it is unlikely that energy consumption patterns will be tamed without significant gov-

---

[26] Majid Shafie-Pour and Mojtaba Ardestani, "Environmental Damage Costs in Iran by the Energy Sector," *Energy Policy*, April 2007, p. 1.

Figure 6.10
The Evolution of Environmental Damage by Sector

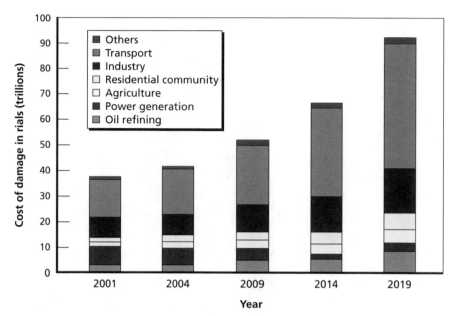

SOURCE: Majid Shafie-Pour and Mojtaba Ardestani, "Environmental Damage Costs in Iran by the Energy Sector," *Energy Policy*, April 2007, p. 9.
RAND *MG729-6.10*

ernment intervention. Urbanization rates in Iran continue to grow (see Figure 6.11), and this will continue to place pressure on the housing and transportation sectors. These factors will lead to growing energy demands.[27]

Iran faces several natural factors that are beyond its control, most notably earthquakes and changing rainfall patterns. The UN reports that Iran is "the number one country in the world for earthquakes—whether measured in intensity, frequency or the number of casualties."[28] Iranian scientists have calculated that Tehran has a 90-percent chance

---

[27] Hamid Davoudpour and Mohammad Sadegh Ahadi, "The Potential for Greenhouse Gases Mitigation in Household Sector of Iran: Cases of Price Reform/Efficiency Improvement and Scenario for 2000–2010," *Energy Policy,* Vol. 34, 2006, p. 42.

[28] Frances Harrison, "Quake Experts Urge Tehran Move," BBC News, March 14, 2005.

Figure 6.11
Iranian Urbanization

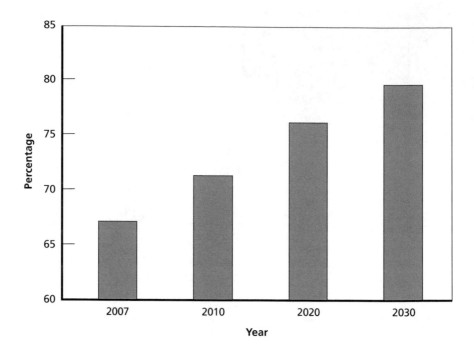

SOURCE: United Nations Population Division, 2004.
RAND *MG729-6.11*

of experiencing an earthquake measuring 6 or higher on the Richter scale and a 50-percent chance of experiencing a 7.5-magnitude earthquake. This has led to recommendations that the capital be moved to a more stable geographic setting.[29]

---

[29] Harrison, 2005. The Iranian Studies Group at MIT notes that

> Iran has been host to a long series of large damaging earthquakes, many of them occurring within the 20th century. There have been roughly 126,000 deaths attributed to 14 earthquakes of magnitude ~7.0 (one 7.0 earthquake/7-yr), and 51 earthquakes of 6.0-6.9 (one/2-yr) that have occurred in Iran since 1900.

See the Iranian Studies Group at MIT, "Earthquake Management in Iran, a Compilation of Literature on Earthquake Management, Draft," January 6, 2004.

In addition to earthquake risk, Iran has been contending with growing rainfall deficits that will affect water availability, water quality, and certain economic sectors (e.g., manufacturing and agriculture). Iran experienced its driest years since 1960 between 1998 and 2000, when precipitation declined by about 43 percent. In more than 270 cities, including Tehran, water scarcity has reached crisis proportions and led to rationing in recent years.[30]

Conclusion

In considering Iran's defensive potential over the near, medium, and far terms, it is worth focusing on factors that Iranian leaders might be able to successfully address. Leaders could conceivably target domestic energy consumption, a problem that was recently tackled through very unpopular rationing measures. Leaders could also initiate steps to increase energy efficiency and improve oil production by investing in modern methods and processes. Leaders could better control state spending through constitutional reforms that move the country toward open markets, reduce the number of SOEs, and encourage development of a more robust private sector that is funded by FDI. Finally, leaders could end Iran's nuclear ambitions and thereby reopen ties to the international community. Most of these changes are unlikely to occur through at least the near term because they would be politically unpopular.

Conversely, the government has little real control over population trends. Although the government has implemented a successful population-control program, the pronatalist policy of the 1980s—and its resulting youth bulge—is a thing of the past that cannot be changed. Iran will continue to face significant pressures on its educational and employment sectors through 2020. Employment strains are not likely to be resolved until there are economic and constitutional reforms that

---

[30] M. Abbaspour and A. Sabetraftar, "Review of Cycles and Indices of Drought and Their Effect on Water Resources, Ecological, Biological, Agricultural, Social and Economical Issues in Iran," *International Journal of Environmental Studies*, Vol. 62, No. 6, December 2005, p. 715.

encourage FDI and allow the economy to create more and better jobs. The government is also not able to control international demand for and prices of oil, a fundamental reason why rentier economies are so vulnerable. If demand and prices remain high through the near and medium terms, the government can put off necessary reforms, but this will only buy the government time. Finally, the government has no control over natural disasters. Iran experiences approximately one earthquake every day, and its current building codes, especially in cities like Tehran, raise the risk of staggering losses of life and property in the event of a catastrophic geological event. The government must be able to maintain the capacity to respond to a natural disaster lest it face eroded popular confidence (as occurred after the Bam earthquake in 2003) and social and economic ramifications.[31]

Iran is unlikely to undertake the substantive constitutional and market reforms needed for job and market growth over the near to medium term because such reforms would be too disruptive. Limited reforms in government subsidy policies are a mixed bag: Though unpopular among the public, such reforms will still be tried because the government must reduce spending. As long as the system has breathing room to absorb the changes, the system will be stable. Ahmadinejad has shown a great willingness to use violence in the last year to control opponents. Only time will tell if this is going to be a successful strategy. It may very well prove successful over the near term, at least as long as he delivers on his populist promise to share oil wealth. With oil prices high, this is more likely. Military ascendancy through the near term will likely continue. Although the functioning of Iran's political system could be expected to improve as underperforming political figures are ousted, the government's violent response to the outcome of the 2006 municipal elections is suggestive of its likely response to similar situations in the near term. How the system will perform during the medium and far terms is unclear.

---

[31] For example, narcotics trafficking and use has increased significantly in Bam since 2003. See Afarin Rahimi Movaghar, Ali Farhoodian, and Reza Rad Goodarzi, "A Qualitative Study of Changes in Demand and Supply of Illicit Drugs and the Related Interventions in Bam During the First Year After the Earthquake," Iranian National Center for Addiction Studies, Winter 2005, pp. 1–28.

For U.S. defense planners, the biggest question is whether the status quo under Ahmadinejad is going to last and what this means for the future. The growing role of the Iranian military is troubling, and it is not clear how societal elements are banding together to confront it. There are signs that Khamenei has been complicit in the military's attempt to amass power; he is probably employing this tactic in an effort to block other political rivals, especially Rafsanjani. If the military assumes a vanguard-of-the-revolution mantel (as did Turkey's military, though with Iran's Islamic revolutionary flavor), then the nexus among Khamenei, Ahmadinejad, and the IRGC could deepen further. The 2008 parliamentary election and the 2009 presidential election will be very important indicators of Iran's potential posture through the medium and far terms. If Ahmadinejad is successful despite worsening economic conditions and growing authoritarianism, the role of the IRGC in particular will have become more politically significant.

In summary, Iran will continue to face persistent economic, social, and political challenges during the medium and far terms. These domestic challenges will constrain Iran's defense spending if it wishes to embark on a sizable military buildup. More likely, Iran will continue to pursue soft power regionally and internationally, a strategy that will not rely heavily on military strength. Iran will most likely continue its support for militias and will help allies (like Hezbollah) gain political power through the ballot box. Iran will also most certainly continue to play a central role in Iraq, at least through the medium term.

CHAPTER SEVEN

Iran's Near Abroad

This chapter examines Iran's near abroad. We begin with a description of the current situation, then assess possible changes in the region that could significantly influence Iranian planning, decisionmaking, and resource allocation. For purposes of this analysis, Iran's near abroad includes not only those nations on its immediate borders, but also countries as far west as Israel.

The Current Situation

Iran is the major near-term beneficiary of the U.S. invasion of Iraq in 2003. The U.S. move to topple the Baathist regime in Iraq removed any near-term (and probably long-term) strategic threat that Iraq may have posed to Iran. Iran's power and influence in the region have expanded significantly since the 2003 invasion; this is especially true of Iran's power and influence over various Shia groups in Iraq.

Although Iran has clearly benefited from the U.S. invasion of Iraq, today's situation is still threatening to Iran. The U.S. invasion unleashed long-simmering Sunni-Shia resentment that has emerged as an escalating sectarian civil war in Iraq. At this time, no one knows how long or how bitter the Sunni-Shia sectarian conflict will be, or whether it will expand to other nations in the region. Already, there is increasing evidence that Iran and its Shia allies in Iraq and Syria are on the other side of an emerging Saudi-led Sunni Arab alliance that includes Jordan, most of the Persian Gulf states, and Egypt. This may lead to a regional Cold War–type conflict in which both sides compete, primarily inside Iraq and Lebanon, through clandestine and covert means. More dangerous is the possibility that internal struggles inside Lebanon and Iraq might ignite an overt inter-Islamic regional war. At a minimum, the existing conflict will likely stimulate sustained arms competition that encourages both sides to attempt to modernize their military forces. In the near term, Saudi Arabia and the members of the Gulf Cooperation Council will exploit their vastly improved financial situation (sustained by high global oil and natural-gas prices) and acquire a new generation of advanced weapon systems. These oil-wealthy states may even subsidize their militarily more-powerful allies (such as Jordan, Egypt, and Pakistan). Although Iran has gained additional revenues from higher oil and gas prices, its mismanaged economy has absorbed much of this windfall by sustaining a wide range of politically popular state-sponsored subsidies.[1]

Iran's barely concealed desire to acquire nuclear weapons has alarmed many nations in the region and raised considerable concern in Europe; it may ultimately even provoke the United States, Israel, or both to take military action against Tehran. At a minimum, Iran risks serious UN-sponsored economic sanctions if it decides to press forward with what may become an increasingly obvious nuclear-weaponprogram.

Highly related to Iran's willingness to press forward with its nuclear program is its relationship with Russia and the PRC. Iran's leaders are very aware that their strategic position will be much more robust if they have some kind of extended security guarantee from Russia, the PRC, or both. Iran's current relations with both nations are in a state of flux. At times, both of the larger nations have seemed willing to support Iran; on other occasions, Russian and Chinese leaders have expressed significant frustration with Iran. Currently, Russia is a major weapons supplier to Tehran. (The recent sale of the SA-15 *Tor* air-defense missile to Iran is an important addition to the Iranian military, for example). China is increasingly interested in obtaining massive quantities of oil and natural gas from Iran under long-term contracts

---

[1] "Geopolitical Diary: U.S., Iran Lose from Major Changes in Baghdad," Stratfor, April 17, 2007.

and therefore has a growing stake in the stability and reliability of Iran. Thus far, neither Russia nor China has been prepared to break diplomatically with the United States, France, and the UK due to the rising tensions over Iran's nuclear program, although the Russians and Chinese were, as of the summer of 2007, clearly not yet willing to support strong UN sanctions on Iran.

Future Trends and Possibilities in Russian and Chinese Support for Iran

As discussed above, political and economic support from Russia, China, or both is an absolutely crucial issue for Iran. Tehran's position would be much more secure if it had the explicit backing of one or both of these nations. The stronger Tehran's ties with Moscow and Beijing become, the bolder Iranian leadership will likely feel in the coming years. As was the case of Cuba in the Cold War, when Fidel Castro had explicit backing from the Soviet Union, Iran's leaders will know that the likelihood of a U.S. or Israeli military move against Iran would be dramatically reduced if the Russians or Chinese made it clear that they supported the Islamic state.

Through the near and medium terms, the PRC's involvement in the region will increase due to China's growing energy needs. Increased interdependence between China and Iran could, in time, lead to a more robust security relationship between the two countries, a relationship that might mimic China's current relationship with Pakistan. Although the Russians do not depend on Iran for energy, Moscow may come to regard a closer link with Tehran as a way of maintaining considerable leverage in the strategic region of the Middle East, especially since Moscow's influence in the Arab world is much weaker than it was in the 1960s through the 1980s. As previously noted, a wealthy and threatened Iran will likely remain a major arms market for Russia. Finally, Moscow has had some success in limiting Tehran's support of Islamic nationalist movements in Russia's near abroad. Specifically, Iran has not yet helped the Chechen rebels (who are radical Sunni Arabs).

In the near term, the "forcing function" that could compel the Russians and Chinese to make fundamental decisions about how strongly they will provide strategic support to Iran is the evolution of the Iranian nuclear-weapon program. If the Iranian choose to defy the UN Security Council (and great powers such as Japan) by pressing forward with its nuclear program, a crisis between the Iranians and the United States or Israel (or both) could arise. At that point, Tehran could press Moscow, Beijing, or both for some kind of security guarantee. In one possible variant of this scenario, China and Russia could transform the SCO into something closer to a political and military alliance and invite Iran to join. This would be a major watershed event in the region and would probably have a very negative effect on U.S.-Russian-Sino relations.

Future Trends and Possibilities in Saudi Arabia and the Gulf Cooperation Council States

In the medium term, Iran and its Arab Shia allies might seek to incite the Shia minority groups (which in many of the Persian Gulf states are a very large minority) inside Saudi Arabia, Bahrain, Kuwait, Qatar, and other gulf states. Leaders of those nations are already concerned that Tehran is doing just that, capitalizing on its stronger position due to the U.S. invasion of Iraq. Should Iran make a serious attempt to inflame and incite the Shia population in these nations, tensions would rise considerably in the entire region, and some form of retaliation on the part of the Sunni Persian Gulf states would probably follow.

In a similar manner, Iran will seek to manipulate the Shia-controlled government of post-Saddam Iraq to serve its own national strategic and economic ends. This will likely alarm both the Sunni Arab nations along the gulf and Jordan.

Iran's pursuit of nuclear weapons could start a very dangerous regional arms race. There have already been reports of Saudi Arabia discussing a possible nuclear program with other Sunni Persian Gulf nations. The most likely route through which Saudi Arabia would acquire a nuclear arsenal and associated means of long-range delivery is cooperation from Pakistan. There is also the precedent of Saudi Arabia's clandestine purchase of CSS-2 intermediate-range ballistic missiles from China during the 1980s. Obviously, Pakistan will think long and hard before considering such a transaction, but both Saudi Arabia and Pakistan may conclude that the Gaullist argument of ensuring sovereignty in a dangerous neighborhood trumps their strategic relations with the United States. Should the Saudis initiate an effort to acquire a nuclear-weapon program, other nations in the region (such as Egypt and Turkey) might feel compelled to do the same.

Israel will certainly find these scenarios strategically appalling. Even if Tel Aviv and Washington do not execute a preventive counter-proliferation campaign against Tehran, the United States may have to offer an extended nuclear-deterrent guarantee to Israel, even though Israel is already nuclear armed. This possible trend toward a larger number of nuclear-armed states in the Middle East would be highly risky and destabilizing. In such circumstances, a new, higher, regional "risk premium" on the price of oil could be a major factor in sustaining high prices for both petroleum and natural gas for the medium and perhaps long term.

Future Trends and Possibilities in Turkey and the Kurdish Independence Movement

Depending on the political outcome in Iraq, the Kurds could press for greater autonomy and (eventually) an independent state. The current situation in Iraq is leaning toward a de facto partition of the country into three quasi-independent regions: Sunnis in the west, north, and center; Shia in the center, east, and south; and Kurds in the north. A Kurdish push for autonomy would alarm both Turkey and Iran. This is already a current and near-term challenge; it could extend into the medium term, since it may take a number of years for Iraq's political fate to become clear. If Iraq does break into three essentially independent regions, the issue of Kurdish independence will become more important. In the near term, the Turkish government might authorize a large-scale use of force to "clean out" radical Kurdish nationalist camps along Turkey's eastern border.

A move toward Kurdish independence might result in a fair degree of cooperation between Ankara and Tehran, since neither is interested in seeing portions of their countries break away to become part of an independent Kurdistan. Today, both the Iranian and Turkish militaries periodically take action against Kurdish extremists inside their own countries' borders. An increased Kurdish push for independence could lead to combined Turkish-Iranian action against the Kurds.

On the other hand, should Iran continue to seek nuclear weapons, there might be a very different reaction in Turkey. The Turks have always considered themselves a major power in the GME. Like Saudi Arabia and Egypt, Turkey might feel compelled to acquire a nuclear capability if Iran becomes nuclear armed. Unlike Saudi Arabia and Egypt, however, it is unlikely that Turkey will be able to acquire its nuclear capability from a foreign source; Ankara would probably have to initiate a "crash" domestic program to develop an independent weapons-grade fissile-material production capacity. Such a profound strategic move would of course prompt an existential crisis in NATO-Turkish and EU-Turkish relations.

Future Trends and Possibilities in Israel, Lebanon, Palestine, and Syria

In the near term, Iran's development of nuclear weapons could provoke Israel to take military action. Such a move on Israel's part would probably receive mixed reviews from Arab states. Certainly there would be no large-scale approval of an Israeli attack on another Muslim nation. On the other hand, several of the Sunni Arab states already feel threatened by the possibility of an emerging Iranian nuclear capability. An Israeli attack on Iran might give several Arab states the best of both worlds: an opportunity to rant at yet another Israeli outrage against a Muslim nation and the quiet satisfaction of seeing the Iranian nuclear program at least temporarily "defanged." It is highly likely that Iran would retaliate against Israeli action in a wide variety of ways, including dramatically escalating its military support to anticoalition forces in Iraq, conducting extensive acts of terrorism, and possibly even shutting down Persian Gulf tanker traffic temporarily.

Beyond the nuclear issue, Iranian support of Hezbollah and Syria is a continued source of concern in Israel and Lebanon. If Iran helps Hezbollah take control of much of Lebanon, Saudi Arabia and its Sunni Arab allies, or the United States or its key European allies, or both, are likely to respond militarily. In the near and medium terms, Iran will probably continue to support Hezbollah while simultaneously expanding its contacts with other radical Islamic groups in Palestine.

In addition to trying to expand its influence by strengthening Hezbollah, Iran will almost certainly continue to improve its relations with Syria. The Syrians believe that Iranian backing helps counter both the U.S. presence in Iraq and the threat posed by Israel. Neither Iran nor Syria approves of the U.S. military presence in Iraq, and both believe that they benefit from the United States' continued troubles there. However, Iran and Syria both hope that the Americans will be forced to leave the region eventually. Due to these shared interests, the Iranian-Syrian linkage will probably continue to grow in the near and medium terms.

Summary

In the near term, Iran has benefited from the U.S. invasion of Iraq in 2003. However, Tehran faces uncertainties associated with the growing Sunni-Shia conflict in Iraq, a conflict that could spread throughout the region. Iran's push for nuclear weapons is also contributing to tensions in the region and beyond.

In the near and medium terms, Iran's near abroad will be a scene of instability and conflict. This turmoil will probably force Iran to devote considerable resources to its military. If Iran elects to increase its efforts to inflame Shia minority groups in the region, or if it is believed to be controlling a puppet Shia government in Baghdad, or if it cracks down on the Kurds in the northwest part of Iran, tension in the region will grow.

From Iran's strategic perspective, much depends on Tehran's long-term relationship with Moscow and Beijing. Obtaining the clear backing of one or both of these nations would greatly strengthen Iran's position.

CHAPTER EIGHT

Japan's Near Abroad

This chapter examines how Japan's relations with its neighbors are evolving. Although the United States is thousands of miles to the east, it is included in Japan's near abroad due to the great importance to Japan of its relationship with the United States.

The Current Situation

Following its defeat in World War II, Japan became a pacifist nation. The postwar Japanese Constitution was essentially imposed on the nation by the United States. Japan's armed forces were reestablished in the 1950s as the Self Defense Force, and strict limits on how and where that force could be employed were established. For decades after World War II, Japan was viewed with considerable resentment and suspicion by many Asian nations that had suffered under Japanese occupation during the 1930s and 1940s.

Three generations after the conclusion of World War II, some Japanese are finally starting to view the country's pacifist constitution as outdated. Japan has begun pushing for high-technology weapons and has begun restructuring its military. Japan now seeks 100 F-22 fighter aircraft, saying the fighter is necessary to combat expansion of the Chinese air force. Japan's acquisition of Patriot Advanced Capability 3 batteries and Standard Missile 3 air-defense systems was in part a response to aggressive North Korean actions.[1]

Commerce facilitated Japan's initial return into the normal community of nations. In the early 1960s, Japan's economy began to expand rapidly, mostly due to Japan's emphasis on exports of steel and other bulk commodities. By the early 1970s, Japan had become a leading player in the global automobile market and was becoming the world leader in the production of high-quality electronic products. Japan's hard-working and highly educated workforce was ideally suited for this export-driven economic strategy. By the mid-1980s, Japan had become a huge economic force in the world, but its strength was built on a bubble economy of stock and real estate. Following the burst of that bubble in the 1990s, Japan's GDP grew at a reduced rate of 1–2 percent per year; this slowdown persisted into the current decade. In September 2007, the Japanese government announced that the economy shrank by 1.2 percent in the previous year.[2] During the post–World War II period, Japan's military role in the world was very limited. Indeed, it was not until Operation Iraqi Freedom in 2003 that Japan deployed troops outside Asia; even then, the Japanese military played only a highly specialized support role. Today, however, Japan is starting to rethink its international role, including the size and mission of its armed forces. The rise of China as a major power in Asia has contributed greatly to Japan's reexamination of its strategic situation.

Current State of U.S.-Japanese Relations

Since 1945, the United States has been the main guarantor of Japan's security. With a constitution that severely limits the role of its Self

---

[1] These actions included Pyongyang's surprise firing of a Taepoedong-1 multistage ballistic missile over Japan in August 1998, the incursion of two North Korean spy ships into Japanese territorial waters in March 1999, a December 2001 firefight between a DPRK spy ship and the Japanese Coast Guard, and North Korea's July 2006 launch of seven ballistic missiles. Soon after North Korea's underground detonation of a nuclear device in October 2006, the Japanese government adopted a number of significant changes to reposture its own military.

[2] "How Fit Is the Panda?" *The Economist*, September 29, 2007.

Defense Force, and with defense spending normally not exceeding 1 percent of GDP between the 1960s and the early 2000s, Japan's military has been regarded as a force that would operate alongside the United States in defense of Japanese national territory alone. Note that although the Japanese have typically spent just 1 percent of their GDP on defense, that percentage has represented a considerable defense budget by most global standards as the Japanese economy has expanded. For example, Japan's 2006 defense spending equaled $41.1 billion. This figure approaches the United Kingdom's 2006 defense budget of $55.1 billion and France's 2006 defense budget of $45.3 billion.[3]

Like many European nations and the United States, Japan is graying. In 2004, 19.5 percent of the Japanese population was over age 65. Japan's current population of roughly 124 million is projected to drop to roughly 100 million by 2050. Japan has never encouraged immigration into the country and is not likely to do so in the foreseeable future. Therefore, Japan's population will become increasingly elderly and the country will become less populous. How these demographic trends will affect Japan's relations with its neighbors is unclear, but the trends strongly suggest that Japan will face an internal labor shortage and increased expenditures for social services.[4]

The past two Japanese administrations have courted closer ties to the United States. Junichiro Koizumi served as Japan's prime minister from 2001 to 2006. His tenure was clearly defined by a strengthening of Japan's political and military relations with the United States. He led his party, the Liberal Democratic Party, to a huge electoral success in 2005. This victory was largely due to economic reforms in the stagnant Japanese economy, including attempts to tackle Japan's increasing government debt. A considerable majority of the Japanese electorate approved of Koizumi's efforts to build improved ties to the United States.[5]

---

[3] See Christopher Langton, *The Military Balance 2007*, The International Institute for Strategic Studies, London: Routledge, 2007.

[4] Japan Ministry of Internal Affairs and Communications, *Statistical Handbook of Japan*, 2006, Chapter Two, "Population."

[5] "U.S.-Japan Defense Alliance Strengthens," *Honolulu Advisor,* June 3, 2007.

During his term, Koizumi upgraded the status of the Self Defense Force by elevating the organization to ministerial level in the Cabinet. When the North Korean nuclear arsenal was exposed, Japan generally backed U.S. efforts to impose increasingly harsh sanctions on the DPRK. Koizumi's successor, Shinzo Abe, was elected in the summer of 2006. During his very brief term in office, Abe continued his predecessor's efforts to increase Japanese assertiveness and tighten relations with the United States. Amid various scandals, Abe was forced to leave office in September 2007. At the time of this writing it is not yet clear how the latest Japanese Prime Minister, Yasuo Fukuda, will approach U.S.-Japanese relations.

In the near term, at least, the U.S.-Japanese relationship will likely remain as close as it has ever been since the end of World War II. U.S. resentment toward imported Japanese products was evident in the United States in the late 1970s and early 1980s, but has since been replaced by an increasing awareness on the part of both countries' elites and general populace that Japanese and U.S. interests are closely related, especially in terms of international terrorism and the evolving strategic situation in Asia.

Japan's Current Relations with Its Asian Neighbors

Japan has been one of the main economic engines in Asia since the 1960s. Japanese companies spread into many Asian locations while Japan was simultaneously targeting the United States as its primary export market. Therefore, many Asian nations that were impoverished at the end of World War II have benefited from Japan's growing economy. It is only in the past decade that China has started to take over as the leading driver of Asia's economy.

On the other hand, much of Asia still feels considerable distrust of and resentment toward Japan. Many countries still remember the rapaciousness of Japan's conquests in the 1930s and 1940s. For example, Prime Minister Koizumi's visit to Japan's Yasukuni Shrine (a memorial to Japanese war dead, where several executed Japanese Class-A war criminals from World War II are interred) led to strong protests from South Korea and China. In the PRC, massive anti-Japanese riots broke out in several cities, clearly showing the extent of resentment that persisted more than half a century after World War II. Many Asian nations feel that Japan has not sufficiently admitted its guilt for the terrors of Japanese imperialism in the 1930s and 1940s; they also feel that Japan should pay reparations for the destruction it caused. Therefore, although many Asian nations have profited from Japan's postwar economic expansion, they remain considerably resentful toward Japan.

Japan's relations with the PRC are of great consequence to U.S. defense planners. The meteoric rise of China—with all the consequences of that rise for Japan—was a major factor in Japan's recent quest for closer ties to the United States. In the near term, therefore, Japan's leaders will continue to believe that China's rise requires closer U.S.-Japanese relations. Although future Japanese governments might think differently, for the foreseeable future it is apparent that Japan regards the expansion of Chinese power and influence with trepidation.

The North Korean situation is also a current and near-term cause of concern for Japan. Much of the bellicose propaganda coming from Pyongyang has been directed against Tokyo. In the event of a second Korean war, all concerned parties (including the DPRK) are aware that the United States will make extensive use of Japanese bases, particularly for air strikes against North Korea. Whereas in prior years the risks to Japan posed by a war on the Korean peninsula were relatively small (e.g., DPRK ships might sink a few Japanese merchant ships, and a relatively small number of high explosive–armed, highly inaccurate North Korean missiles could strike inhabited areas close to U.S. bases), today the situation is very different. The DPRK's acquisition of nuclear weapons has profoundly raised the stakes for Japan. Thus, the twin events of the rise of China (which is still suspicious and distrustful of Japan) and an unstable—and now nuclear-armed—North Korea have made Japan's strategic situation less safe.

The renewed flurry of diplomatic activity during the summer of 2007 suggested that the DPRK may protract the process of disarming its nuclear-weapon infrastructure. However, Pyongyang has shut down a small plutonium-producing reactor at Yongbyon and allowed IAEA inspectors to confirm this closure. The IAEA has also confirmed that construction of the nearly completed larger reactor has ceased. Finally, Pyongyang announced that Kim Jong Il will meet with South Korean President Roh Moo-hyun in late August 2008 for a second summit. These events suggest that North Korean leadership may have decided to abandon the country's nuclear arsenal. On the other hand, these decisions may be part of North Korea's new "peace campaign," which is designed to weaken the rising political prospects of the more-nationalist and more-conservative political forces of South Korea's Grand National Party during the upcoming South Korean presidential election campaign in December 2008. At present, Roh's Uri Party appears to be in serious political trouble. Put simply, North Korea may not have given up the hope that it can maintain a small nuclear arsenal without paying a meaningful economic or diplomatic price. How this drama unfolds will have a strong effect on Japanese elite and public attitudes toward a wide range of emerging national security issues, and will color Japan's relations with China.

Japan is also losing its decades-long economic leadership position among Asian nations. Whereas Japan was clearly the dominant Asian economy from the 1960s to the 1990s, today China has taken that place. Chinese labor is much cheaper than Japanese labor, although Chinese workers are at present far less skilled as a group. Over the past decade, China has made great economic gains among Asian nations, often taking over from Japan the role of principal regional trading partner with various nations. As of 2006, China's GDP was roughly $2.6 trillion, while Japan's GDP was about $4.7 trillion. In terms of per capita wealth, Japan is far ahead of the PRC (Japan's per capita GDP is $34,000 per capita; China's is $2,000). Nevertheless, if China's projected growth rates prove accurate, the PRC will probably overtake Japan's total GDP by late in the current decade or during the early years of the next. Of course, much of China's economic surge is due to continued expansion of Chinese exports in Asia and elsewhere. To some extent, this Chinese expansion has come at Japan's expense.

Japan maintains generally cordial relations with other Asian nations, although none of these relationships involves binding security agreements. Japan is an active member of the UN and the East Asia Summit. Japan also donates substantial sums to international assistance efforts. Therefore, Japan has made its presence felt on the international scene, especially since the 1970s, but in a very nonmilitary way. Today, Japan is regarded by most Asian nations as a strong economy that is, however, aging and gradually losing its position of leadership to a rising China.

Future Trends and Possibilities

Japan is now watching very important changes take place in Asia, mostly due to the rise of China. How Sino-Japanese relations evolve over the next decade or two will have profound impact on Japan. Whereas today the Japanese Self Defense Force is not explicitly config- ured to contest the PRC, should the two nations come to regard each other as clear threats, that would almost certainly change.[6]

Japan has recently had to place considerably greater emphasis on missile defense compared to previous years. Whereas the North Korean threat used to consist of inaccurate, conventionally or chemi- cally armed missiles, Japan now lives under the shadow of DPRK's nuclear weapons. Until and unless the DPRK threat is neutralized (via regime "implosion," reunification of South Korea, or a denucleariza- tion treaty), the Japanese will almost certainly continue to gradually improve their defenses.

Regarding missile defense, the Japanese may face the issue of scale if China comes to be regarded as a significant threat. Whereas relatively modest Japanese (or U.S.-Japanese) missile defenses might be able to protect most of the country from the limited North Korean missile capability, a much larger and more robust defense system would be required against China.

If Sino-U.S. relations deteriorate in the future (over the Taiwan issue, for example), the Japanese will face a difficult situation. Whereas the stakes for Japan in a second Korean war are relatively limited (although North Korea's acquisition of nuclear weapons has certainly increased the risk to Japan), a major Asian war involving the United

---

[6] "The Hitch in Japan's Normalization Plan," Stratfor, September 12, 2007.

States and China could have far graver consequences for the Japanese. The warning signs of worsening Sino-Japanese relations would probably make themselves clear over a period of years. During that time, the Japanese would probably come under increasing pressure from both the United States and China. The former would want assurances from the Japanese that its bases would be available for U.S. use in the event of a showdown with China; the United States might make other demands of Japan as well. The PRC, on the other hand, would probably make it known that it would regard any American use of Japanese bases to strike China as something akin to an act of war on Japan's part, whether Japanese forces participated in such strikes or not. Currently, U.S. bases on Okinawa are the closest to the PRC and would figure prominently in any calculation on Japan's part about whether it should risk supporting the United States in the event of a war with China. Such risk would be considerable. Even today, the PRC could hammer Okinawa with a large missile barrage; in the near to medium term, the Chinese ability to hit Okinawa will improve dramatically as the range, accuracy, and number of Chinese medium- and intermediate-range ballistic and cruise missiles continue to grow. Farther in the future, the defense of military targets on Okinawa from large-scale and accurate conventional-missile attacks from China will require a major investment in active and passive defense measures—an investment that Japan is not prepared to make.

The main point is that as Sino-U.S. relations in Asia evolve in the coming years, Japan will be greatly affected. It will probably be pressured by both the United States and China to "take sides." Should the Japanese perceive the threats to them as being increasingly severe, the country might take the additional step of wanting to acquire its own nuclear arsenal.

In the medium to far term, the situation on the Korean peninsula will also change compared to today. As noted above, from Japan's point of view, a "soft landing" regarding the two Koreas is highly desirable. A worst case for the Japanese is a second Korean war in which nuclear weapons are used. This scenario includes the possibility of North Korean nuclear strikes against Japan or of fallout coming across the Sea of Japan. Therefore, the Japanese will probably attempt to strike a balance between being "tough" on the regime in Pyongyang as long as it retains nuclear arms while at the same time not pushing so hard that it backs the North into a corner from which it comes out swinging. The Japanese goal will be to gradually help bring the Korean problem to a peaceful conclusion. Once that happens, however, other issues will come to the surface.

In the event of a reunified Korea, and even if the reunification occurs peacefully, Japan will face the prospect of a nuclear-armed Korean state of uncertain intent. As mentioned earlier, many Asian nations still harbor ill feelings toward Japan. The Koreas are certainly among this group. It is not clear whether a unified Korea would align itself with the PRC or with the United States and Japan. If in the medium to far term Japan comes to regard Korea as being unfriendly, there would be increased incentive for Japan to (1) increase its own defense capabilities and (2) further strengthen its ties to the United States.

Although the evolution of China and the Korean peninsula will be of central concern to both the Japanese elite and the general public, the evolution of Tokyo's relations with Moscow will also be significant. Russia may become a major supplier of both natural gas and oil to Japan's reviving economy. The nearby Sakhalin Island gas and oil fields may make the Russian Federation a very attractive energy supplier of choice. In spite of its robust (and occasionally troubled) nuclear power program, Japan remains very dependent on the steady flow of large quantities of oil from the Persian Gulf. Liquefied natural gas from the Middle East may become another key energy source for Japan.

At the strategic level, Japan has three broad options available to it in the coming years.

Continue to strengthen its relationship with the United States. This is Japan's present course of action. The rationale supporting such a strategy is that as China's power and influence rise in Asia and elsewhere, the Japanese should counterbalance those effects with an increasingly "special" relationship with the United States, similar to the special relationship between the United States and the United Kingdom. In this scenario, Japan's military ties with the United States will continue to grow. Of course, the PRC will probably see this as unfavorable to its interests.

Reach an accommodation with China. The second possibility is that Japan will conclude that U.S. power and influence in Asia is declining and that China's will continue to rise in the coming decades. In this case, Japan will seek to reach an understanding with the PRC regarding Japan's "place" in Asia. Japan would remain a major economic power, but its foreign policy would become increasingly accommodating to China's will. The United States would be the primary loser in this scenario.

Pursue a more independent foreign policy. In this scenario, the Japanese will, like the French in the 1960s, conclude that they cannot continue to be as dependent on U.S. protection for their security. They will thus decide to increase their own military power independent of American plans. This scenario could ultimately include Japan's acquisition of nuclear weapons.

CHAPTER NINE

Russia's Near Abroad

This chapter discusses the status of Russia's current domestic situation and its relations with its neighbors. First we examine the current state of the Russian economy, the role of hydrocarbons as drivers of present (and future) Russian growth, and current developments in domestic politics. Then we explore Russia's relations with Europe, China, and Japan. Finally, we review future challenges and potential impediments—most of which are domestically rooted—to further economic development and their implications for Russia's relations with its near abroad.

The Current Situation

Russia has recovered from the 1998 financial crash that caused it to devaluate the ruble and to default on its foreign debt commitment. Russian aggregate GDP grew an average of 6.1 percent per year in real terms between 2000 and 2005. Because of falling population, per capita GDP grew even more rapidly (at 6.6 percent each year). According to the International Monetary Fund, Russian GDP hit $1 trillion in 2006, which puts the country back in the group of the world's ten-largest economies. Despite its rapid growth, however, Russia is still a small economy compared to that of the United States (see Table 9.1). In 2006, the Russian economy was the 14th-largest economy and amounted to one-thirteenth (less than 8 percent) of the U.S. economy.

These figures are somewhat different when the comparison is done on a PPP basis, in which data are adjusted to reflect the generally lower

Table 9.1
Gross Domestic Product in 2006

Metric	Total Value ($ billions)	Share of U.S. Value (percent)
Aggregate GDP		
United States	13,194.70	N/A
Russia	984.93	7.5
China	2,644.64	20.0
Per capita GDP		
United States	45,593.85	N/A
Russia	6,897.23	15.1
China	2,012.52	4.4

SOURCE: International Monetary Fund, World Economic Outlook Database, 2006a.

NOTES: Foreign currencies are converted into U.S. dollars at market exchange rates. Annual growth reflects growth of real GDP in local currency units.

prices in poorer countries. On this basis, Russia's economy amounted to 13.4 percent of the U.S. economy.

Hydrocarbons are a major factor in the Russian economy. Russia owns 26.6 percent of the world's proven gas reserves and 6.2 percent of the world's proven oil reserves. In 2005, the country accounted for 21.6 percent of global gas production and 12.1 percent of global crude production.[1] Oil and gas contributed to about 20 percent of Russian GDP.[2] Counter to conventional wisdom, however, the driver of Russia's six years of economic growth has mainly been a boom in domestic consumption, not growth in the energy sector. Especially in the financial sector, construction and trade have outperformed other sectors and have grown above the country's average.[3] After the 1998 meltdown of

---

[1]   British Petroleum, *Statistical Review of World Energy*, London, 2006.

[2]   World Bank, *Russian Economic Report No. 13*, Washington, D.C., 2006.

[3]   Organisation for Economic Co-operation and Development, *OECD Economic Surveys: Russian Federation*, Vol. 2006, No. 17, Paris, 2006.

the Russian economy and its state budget, the ruble was devalued, government spending was cut, and a low flat tax was instituted in 2000. The devaluation led to greater competitiveness of Russian products overseas and kick-started domestic growth again. Rising oil and gas prices also supported economic recovery, as did general improvements in management and technology in private companies.

Nonetheless, the bulk of Russian state revenues stems from hydrocarbon sales. In fact, the recent oil-price increases have resulted in soaring tax revenues. The total share of government revenues from oil and gas (including taxes on refined products, royalties, export taxes, and profit taxes on oil and gas companies) has more than doubled during the last four years, amounting to almost 40 percent in 2006.[4] Energy revenues are much less dominant in financing governmental expenditures than revenues would suggest, since more than half of the oil and gas revenues are saved in a "Stabilization Fund"—a lesson learned from the 1998 plunge in oil prices and its devastating effects on state finances. Hence, energy revenues in 2006 financed only about one-fifth of total government expenditures.

The Kremlin's policy since early 2000 has generally aimed to strengthen governmental control over the national economy. Although this trend has been met with concerns in Western countries and the private sector, the Russian population, having experienced a chaotic political and economic decade after the collapse of the Soviet Union, largely welcomed resurging state power. Especially with regard to the "oligarchs," a small group of elites who had managed to grab the bulk of the state's assets during the shady privatization process of the 1990s, the electorate largely approved Putin's proclaimed "dictatorship of law." During eight years of President Putin's reign, Russia clearly gained increased internal strength and cohesion. Now, the Russian "state" is strong again, a remarkable shift from the 1990s, during which tumbling state authorities sparked debates of Russia as a potentially failing state. In line with this general trend to regain control of the economy,

---

[4] International Monetary Fund, *Russian Federation: 2006 Article IV Consultation—Staff Report; Staff Statement; and Public Information Notice on the Executive Board Discussion*, Country Report No. 06/429, Washington, D.C., December 2006b, p. 35.

the Russian government has gradually restricted foreign ownership in "strategic sectors" of the Russian economy (i.e., the aerospace, military and nuclear power, and hydrocarbon extraction sectors). Two recently approved laws sharply restrict foreign ownership of oil and natural gas assets. After state-controlled Rosneft's 2004 takeover of Yuganskneftegas (formerly Yukos' cash cow) and Gazprom's $13 billion purchase of Sibneft (now Gazprom-Neft) in 2005, the Russian state now controls a significantly higher share of the domestic oil industry than it did during the 1990s. At present, state-controlled companies contribute around 25 percent to the country's oil production and hold around 16 percent of its refining capacity.[5] In the gas sector, state-controlled Gazprom accounts for approximately 86 percent of Russian gas production.

However, the Russian oil and gas sectors face heavy investment challenges. As major oil fields have peaked, the development of new fields will have to compensate for declining production and cover all of Russia's planned annual output growth. The Russian Ministry of Energy estimates that necessary accumulated investment in the oil sector until 2020 may reach $240 billion. The IEA's estimate through 2030 is $400 billion.[6] As for gas, major producing fields are already in decline; this includes the "Big Three" (Yamburg, Urengoy, and Medvezh'ye), which currently account for more than 60 percent of total Russian production. Because most of Russia's known reserves are in the far north, costs for exploration and production (E&P) of new fields will rise significantly in the future due to difficult geological conditions and the arctic climate. According to IEA estimates, Gazprom will have to spend an average of $17 billion per year, which totals more than $400 billion, through 2030 to finance gas-related E&P projects and maintenance of current fields.[7] Meeting these investment challenges will be

[5] See Yulia Woodruff, "Russian Oil Industry Between State and Market," in *Fundamentals of the Global Oil and Gas Industry*, 2006, Petroleum Economist, October 2006.

[6] Russian Ministry of Energy, *Energy Strategy of Russia for up to 2020*, Moscow, 2003; and International Energy Agency, 2004.

[7] International Energy Agency, 2006, p. 123. The Russian Energy Strategy forecasts investment needs of $120 billion to $200 billion between 2002 and 2020. See Russian Ministry of Energy, 2004; and Vladimir Ivanov, *Russian Energy Strategy 2020: Balancing Europe with the Asia-Pacific Region*, ERINA Report, Vol. 53, 2003.

crucial in securing a steady financial return for the Russian state from the extracting industry.

Despite its resource wealth, Russia has apparently managed to escape one curse of resource-based economies, which is to build in recurring government expenditures based on high resource prices and then to have to retrench when those prices inevitably fall. In fact, sound macroeconomic policy has accompanied Russia's economic recovery and expansion. Fuelled by running trade and current account surpluses as well as a budget surplus, the Stabilization Fund prepares for the day that resource revenues fall. Some economists believe that Russia could get an extra economic boost by investing part of the Stabilization Fund in developed-market equities. Moreover, some observers note, the same private, resource-based conglomerates that led the recovery of oil and gas plowed their revenues into manufacturing companies and restructured them, acting much as Asian trading companies have in the past. The services sectors have grown as well, spurred by an increasingly wealthy population. Given Russia's strong base in oil, gas, timber, minerals, and other resources, it is conceivable that resource-based manufacturing could continue to grow over the long term. This is especially true because of Russia's still-high education levels, which some economic analyses have shown are helpful to the development of resource-based economies. Highly educated Russians could also provide the base for developing high-technology capital goods industries, such as machinery, equipment, aircraft, and rockets. Should these trends continue, incomes will remain high and the services sectors will remain strong.

In line with the country's recent economic recovery, the Russian defense budget is rapidly increasing. It is difficult to estimate the exact amount of Russian defense spending, but the 2007 Russian military budget totaled approximately $30 billion (about 2.5 percent of GDP). Defense spending has increased 500 percent since 2000 and is expected to further increase in the future by double-digit annual growth rates.[8] Despite high growth rates, however, overall Russian defense spending is still small compared to Western European countries.[9]

---

[8]   Bank of Finland, *BOFIT Weekly*, Vol. 28, July 12, 2007.

[9]   Langton, 2007.

Russia's strengthened internal cohesion and its regained economic strength enable Russian leaders to take up distinct positions in foreign policy, translate interests into foreign-policy actions, and draw on resources that were not at their disposal a decade ago. Russia is pushing its claims in Central Asia, forcefully opposing the installation of the planned U.S. missile-defense system in Eastern Europe, and demonstrating revived influence in Southern Europe and the Balkans. In addition, it is currently opening new foreign-policy fronts by taking an early lead in the emerging race for the Arctic, which some experts believe contains up to 25 percent of the world's yet-undiscovered oil and gas reserves. As the polar ice cap shrinks, the Arctic region will become more accessible and thus be subject to rival claims among the eight countries with Arctic borders—including the United States.[10] Russia claims that the underwater Lomonosov Ridge is an extension of the Russian landmass, rendering a major part of the Arctic region Russian territory under the UN Law of the Sea Convention. Although the Arctic-bordering countries' high hopes for Arctic energy wealth may never be fulfilled, the emerging "Arctic gamble" may force the Russians to dedicate a higher share of the country's civil and military spending to securing its interests in the Arctic region.

Russia's domestic political situation has become fairly stable after eight years of President Putin in power. His "managed democracy" combines a strong executive and enhanced state control of the media and society with an economic system based on market principles. After the 2003 elections (reconfirmed by recent 2008 elections) and major amendments to Russian electoral law, United Russia, a Kremlin-steered center-right party, controls the majority of the State Duma. Putin's successor, new Russian President Dmitri Medvedev, was handpicked by Putin and elected with more than 70 percent of all votes. Medvedev, member of Putin's St. Petersburg "Siloviki" group and his long-term political comrade, has served almost 17 years under Putin. Lately, he was the First Deputy Prime Minister and chairman of Gaz-

---

[10] The eight bordering countries are the United States, Russia, Canada, Iceland, Norway, Sweden, Finland, and Denmark.

prom. Hence, continuity can be expected in terms of both personnel and general policy.

Remarkably, as the end of his second and final term approached, President Putin managed to avoid being branded as a lame duck in domestic politics. After long and extensive speculations about his political future, and given the former president's persistent popularity, Putin has announced that he will become Prime Minister of a United Russia–led government. Hence, he will certainly retain a powerful domestic position that ensures his influence on future developments in Russian politics.

Current Russian-European Relations

The EU-Russia Partnership and Cooperation Agreement (PCA) forms the backbone of the EU-Russian relationship. The PCA, signed in 1994 and ratified in 1997, sets the regulatory framework for the political, economic, and cultural relations between both parties. It also provides the legal basis for bilateral trade. The EU is Russia's most important trading partner, accounting for about 52 percent of the country's overall trade. Russia, in turn, is the EU's third-largest trading partner, accounting for 8.3 percent of European foreign trade. Bilateral trade relations are asymmetric, though, since Russian exports to Europe mainly consist of energy and minerals (65 percent), whereas the EU's exports to Russia are diversified and mainly consist of industrial and manufactured goods and chemicals. As of 2007, more than 50 percent of Russia's oil exports and more than 60 percent of Russia's natural gas exports go to the EU. In 2007, Russia was the EU's most important supplier of oil and gas, accounting for approximately 50 percent of European gas imports and 25 percent of its oil imports.[11] In turn, and as a consequence of a highly regulated domestic Russian gas market, Gazprom earns virtually all its profits from exports to Western Europe,

---

[11] European Commission (DG Trade), *EU Bilateral Trade and Trade with the World*, 2007; and European Commission, *Russian Federation—Country Strategy Paper 2007–2013*, undated.

although this market only accounts for less than 30 percent of total Russian production.[12]

Russia's recent gas disputes with Georgia, Ukraine, and Belarus have raised fears among the Europeans that Russia could use energy as a foreign-policy instrument. Moreover, current Russian attempts to vertically integrate into the European downstream sector and to curb alternative supply routes to the European gas market by requiring Central Asian producers to ship their gas through existing Russian pipelines have raised serious concerns. The EU has therefore initiated many initiatives aimed at increasing security in European-Russian energy relations. These initiatives range from the—largely unsuccessful—Russian-European Energy Dialogue to a strong emphasis on energy issues in ongoing renegotiations of the PCA. The recent U.S. initiative to install a missile-defense system in East European member states further deteriorated Russian-European relations and led to harsh Russian reactions, including Russia's withdrawal from the Treaty on Conventional Forces in Europe and President Putin's announcement to target Europe with nuclear ballistic or cruise missiles again. In addition, repeated European criticisms of Putin policies that disrespect democratic values and undermine the principles of the EU-Russian partnership, as well as Russia's increasing tensions with former satellites (such as the Balkans), have fueled an atmosphere of mutual Russian-European distrust. Finally, Ukraine and Georgia regard the EU's active "Neighborhood Policy" as a gateway for EU accession. After several waves of EU enlargement have steadily and consecutively restricted Russian influence in Eastern Europe, Russia perceives potential further eastward enlargements of the EU as a threat to its economic and security interests in its near abroad.

It is important to note that Russia does in fact not have an "energy weapon" in oil or gas. Russian oil companies—both state-controlled and private—trade most of their crude on a globalized market, which levels off any single producer's leverage over consumers. Unless a majority of Russian crude is tied up in bilateral contracts, a rather unlikely scenario in the foreseeable future, Russian oil companies will not gain

---

[12] Gazprom, *Annual Financial Report 2006*, Moscow, 2007.

any leverage over individual consumers. As for gas, which is almost exclusively transported via pipelines, long-term bilateral agreements dominate contractual relations. Hence, given extremely high upfront costs for a gas-pipeline grid, it becomes very costly for either involved party to leave an established, bilateral, contractual gas relationship. This mutual dependency renders unilateral action financially painful for the producer. In addition, the emergence of liquefied natural gas will contribute to the development of a global gas market that essentially mimics the oil market.

Current Sino-Russian Relations

Traditionally strained bilateral relations between Russia and China have experienced considerable improvements in recent years. In 2005, both countries solved remaining border disputes over territories in Russia's far east. They also held several joint military exercises in 2005 and 2007, the last of which—"Peace Mission–2007"—took place in China's Xinjiang region and then in Russia's Ural Mountains. Moreover, China and Russia collaborate through the SCO, which has become a major vehicle to limit U.S. (and EU) influence in both countries' near abroad. In 2006, China was Russia's fourth-largest trading partner; Russia emerged as China's eighth-largest trading partner. Again, mutual trade relations are asymmetrical. Russian exports to China mainly consist of natural resources and raw materials. Overall, oil and petroleum products account for nearly 50 percent of Russian exports to China. With the exception of military sales, the share of natural resources in Russian exports to China has tended to increase, whereas the share of mechanical and industrial products has tended to decrease. Despite that trend, however, Russia accounts for only 11 percent of China's overall imports in oil and oil products; whereas the bulk of Chinese oil imports stem from Saudi Arabia, Iran, and Angola. Chinese exports to Russia, in turn, are increasingly diverse. In 2006, machines and technical equipment made up 29 percent of China's exports to Russia, followed by textiles and electronic devices.

Present Sino-Russian collaborations—within the SCO and beyond—seem to serve the countries' mutual interest in balancing U.S. dominance in strategic neighboring regions. At the same time, there are strong limits to Sino-Russian cooperation, especially with regard to increasing competition over Central Asian energy resources. Russia's recent and successful efforts to enhance control over Central Asian gas by requiring Turkmenistan to ship its gas to Western markets via Kazakhstan and Russia was not only a blow to U.S. and European efforts to tap energy sources independent of Russian influence; it was also a step toward fiercer Sino-Russian competition for regional energy resources. China's recent oil and gas deals with Central Asian countries, in turn, include a 30-year natural gas contract with Turkmenistan of 30 billion cubic meters annually and an oil-pipeline project with Kazakhstan that constitutes 1 million barrels per day. These arrangements provide China with access to the Caspian Sea's rich oil resources. However, the arrangements have thwarted Russia's strategic goal of monopolizing Central Asian gas and have deprived Gazprom of an indispensable fallback if domestic production does not keep pace with increases in domestic demand and export commitments. In line with this strategic competition, several high-profile energy contracts between Russia and China, including the much debated Altai gas pipeline, have been put on hold.

Although remote at this time, it is possible that Russia and China could significantly strengthen their geostrategic and military ties. This move would likely stem from a serious downward trend in each country's relations with the United States. This downturn could be prompted in large part by increased tension with the United States over each country's near abroad (for example, Ukraine and Taiwan).

Current Russian-Iranian Relations

Iran is among Russia's most important Middle Eastern allies. Russian-Iranian relations are grounded in a quid pro quo: In exchange for taking a pro-Iran stance in the UN, Russia is granted a free hand in the Caucasus, a predominantly Muslim region. Both countries share important mutual interests in Central Asia, most prominently balancing U.S. influence in the region. Furthermore, Moscow needs Tehran's compliance in yet-unsettled disputes over the Caspian Sea's rich energy resources. Finally, Iran is one of the most prominent supporters of the planned establishment of a cartel of gas exporting countries (the so called Gas-OPEC consortium). Russia has advanced this concept to strengthen the Russian position in the world's gas markets, but Iran regards such a cartel as an opportunity to contain Western influence and power.

Besides these geopolitical considerations, however, there is comparatively little cooperation of both countries in the energy sector. This is astonishing because Russia and Iran together account for almost one-half of the world's oil and gas resources. Russian oil and gas companies are not prominently present in Iranian upstream projects, despite the striking lack of foreign investment in Persian oil and gas. However, Moscow has played a crucial role in Tehran's nuclear program by constructing the contested light-water nuclear reactor at Bushehr. In addition, Russia is one of Iran's most prominent suppliers of weaponry. It has sold a variety of major weapon systems to Iran, including tanks, missiles, and surface-to-air missile-defense systems. Finally, with the support of Russia, Iran gained observer status in the SCO in 2005, an organization that is largely regarded as a Russian-Chinese vehicle to constrain U.S. influence in Central Asia.

Despite Russia's commercial interests in Iran and the two countries' shared geostrategic goals, Moscow is not interested in having a nuclear-armed country in its southern backyard, and it does not support nuclear proliferation in the Middle East. As several Russian observers have noted, Moscow lies well in the range of current Iranian missile systems, whereas the European capitals do not (yet). Moreover, despite recent Russian-Iranian rapprochements, mutual relations have overall soured during recent years, mainly due to Tehran's increasingly elusive policy toward the global community. Hence, as the recent Russian rejection of full Iranian SCO membership and the slowdown of construction activities at Bushehr suggest, Moscow increasingly seeks to keep a cautious distance from Tehran. Given Moscow's simultaneous interest in maintaining political leverage over Tehran, Russia is forced to make a sharp diplomatic tightrope walk in order to not fully alienate the West. In all, however, and despite present political backing in the UN, Iran will not gain an extended deterrence commitment from Russia in the future.

Current Russian-Japanese Relations

Russia's relations with Japan are hampered by a dispute over territorial ownership of the Southern Kurile Islands, which were annexed by the Soviet Union at the end of World War II. To date, the dispute has prevented both countries from agreeing to a formal peace treaty. This political stalemate is reflected in grossly underdeveloped economic ties between the countries. In 2005, foreign-trade turnover between Russia and Japan amounted to $10 billion—a fraction of trade volumes of both countries with third parties.

Despite the Kurile issue, however, Japan and Russia have recently made progress in developing their relationship. Out of the six main areas of mutual relations both countries had defined in 2003, economic relations—and particularly energy sales—prove to be Russia's main interest. This interest meets rising Japanese concerns about its own energy security. Japan, which presently imports 89 percent of its oil from the Middle East, needs Russian energy to enhance the security of its supply. In fact, Japan's national energy strategy favors "resource diplomacy" aimed at deepening relations with energy-supplying countries.

Recently, Russia signed several energy deals with Japan, including a preliminary agreement on long-term deliveries of liquefied natural gas. Despite Japan's urge for diversification of its oil imports and its growing need for (Russian) natural gas, Japanese reservations about Russia may hinder a rapid rapprochement of both countries. One reason is that Moscow has for years played Tokyo against Beijing over the planned 1.6 million-barrel-per-day Eastern Oil Pipeline that will connect the Siberian Angarsk oil fields with either the Chinese or the Japanese market. Although Russia has finally opted for Japan and has committed to spend more than $20 billion to build the pipeline to the Pacific coast, Japan remains suspicious about Russia's reliability in energy issues. In addition, Japanese companies—along with Western firms—were recently forced to considerably diminish their involvement in the Sakhalin-2 liquefied natural gas project that is vital to Japanese energy security.

Future Trends and Possibilities

By 2020, Russia will likely emerge as a successful high-technology petro-state. Domestically, President Putin's political heritage will bring about both positive and negative results. Having further strengthened the executive and having led Russia back to budget surpluses, Putin's and his successors' credentials will lie in having brought a measure of stability to the Russian system and in having reestablished a sound and future-oriented macroeconomic management policy for the country. At the same time, however, the domestic trends toward enhanced state control of society, nationalism as the overall dominant political narrative, and an authoritarian type of state capitalism will have gained further momentum and will have become the dominant characteristics of the Russian state.

Hence, Europe remains disturbed by Russia's failure to become a democracy, and tension over the political and geostrategic fate of Ukraine, Belarus, and Georgia persists. In Central Asia, Russia will keep much of its Soviet-era influence because of traditional political and economic ties, a persistent linguistic advantage, and still-strong Russian soft power. The region will, however, become an increased source of competition between Russia and China.

The Russian economy will probably continue to grow between 5 percent and 6 percent per year because of high energy prices and the development of a high-technology capital-goods sector (that includes globally competitive companies in selected sectors, such as machinery, military technology, and aerospace). If the United States maintains 2000–2005 economic growth rates, then by 2025, the Russian economy will be about 12 percent of the U.S. economy when valued at market rates (at PPP rates, the percentage would be larger). The Russian economy will maintain extensive trade ties with Europe and will be globally integrated.

The evolution of the global energy balance, e.g., the global dependence on the flow of petroleum and natural gas from the Persian Gulf region, may have profound geostrategic consequences. If the United States and its major industrial allies conduct a sustained effort in energy diversification, such as developing biofuels and pursuing energy conservation, then the global economic exposure to a major petroleum or natural gas disruption in the GME will be reduced. Furthermore, a soft energy market could lead to a major and sustained fall in the price of petroleum and natural gas sometime between 2010 and 2020.

A soft energy market could also affect Russia. Not unlike the Soviet Union's problems during the collapse of petroleum prices in the early 1980s, the Russian Federation might face acute financial, economic, and social stress if the global energy market turns soft. Such internal strain might well limit Moscow's willingness to subordinate its geostrategic and geoeconomic interests in the name of a Sino-Russian alliance.

Hence, there will likely still be limited technological and military cooperation between Russia and China. This relationship will be based on Russia's desire to preserve a strategic advantage with still-formidable military-industrial capabilities, even with China's growing fiscal and technological resources. Put simply, the behavior of China and Russia on the military dimension will be determined on a case-by-case basis, with neither side fully committed to institutionalizing this relationship of strategic cooperation.

A successful Russian economy also means a Russia that is more able to modernize and expand its military. In particular, looking out to mid-2020, it is plausible that advanced-technology industries (such as aerospace, precision machine tools, and communications) could expand, improving the capabilities of the Russian military. Furthermore, Russia is likely to remain a very aggressive exporter of advanced weapon systems and military-industrial technology, especially to countries that may challenge U.S. geostrategic interests. Obvious candidates include China, India, Iran, Syria, and Venezuela. The motive for many of these sales will be to generate income for the Russian aerospace sector. Other motives will include more-traditional "loss leaders" to gain or sustain geostrategic influence.

Much could go wrong for Russia, however. The biggest challenge it faces through 2025 and beyond is its demographics. Russia's overall population is expected to decline, but its working-age population, which peaked in 2000, is expected to decline even faster. With the decline of Russia's young and working-age populations, the country will experience a dramatic rise in the dependency ratio: It will increase from 0.41 in 2005 to 0.50 in 2025. The decline of the number of working-age Russians will hit Russian men especially hard.

These trends will likely have little effect on industries (such as high-technology industries or capital-intensive resource industries) that use highly skilled workers. To maintain strong growth, however, Russia may have to tolerate sizable immigration flows.

The flip side of these demographic trends is that, as eastern Russia especially becomes depopulated, another population will move in—the Chinese. This prospect is already raising concerns among Russian industrialists and government officials. Less than 200 years ago, eastern Russia, with its vast stock of resources, belonged to China. Over the next two decades, Russia may face a security dilemma regarding its eastern population and territory while it simultaneously builds economic and security ties to a growing China.

There are also indications that Russia's economic policies and governance could turn counterproductive, hurting its growth prospects through 2025. Several problems have already emerged. The most important is that although Russian economic growth was based mostly on the efforts of private-sector companies, there has been a slow trend toward renationalization, with SOEs buying the more dynamic private companies.[13] In fact, the trend toward state control of "strategic sectors" suggests that Russia may be adopting a Colbertist-type model. Between 2004 and 2005, the private sector shrunk from 70 percent to

---

[13] Former Putin Economic Advisor Andrei Illariounov cites Yaganskneftegaz, Sibneft, Silovye Mashiny, Kamov, OMZ, and Avtovaz as examples of major industrial companies taken over by state-owned companies (Andrei Illariounov, "Russia Inc.," *New York Times*, February 4, 2006).

65 percent of the economy. An enhanced state grip on the private sector has to be expected.

Much depends on Russia's ability to tackle the lack of infrastructure investment, a difficult regulatory environment for growing (rather than starting) a business, and corruption. Given its high share of hydrocarbons and raw materials in its exports, Russia must also address the risk of suffering from the resource curse. The economic side of this phenomenon, dubbed "Dutch disease," may endanger the country's overall economic performance. When a country experiences a resource boom, its income rises and its demand for nontradables (domestically produced goods that do not compete on the world market) goes up. This leads to an appreciation of the real exchange rate, hurting exporters and shrinking the manufacturing sector. Since manufacturing is thought to provide greater opportunity for productivity gains than nonmanufacturing, although the country is richer, its growth prospects are worse. The political effect of resource wealth is that it tends to render a central government more independent of its electorate as a growing percentage of state income comes from taxing natural resources. This leads to a decline in democracy and, again, increased corruption.

Overall, expanding state ownership of the economy, a decline in manufacturing, and a declining population might not harm Russia's military capabilities, especially those that rely on technology rather than personnel. However, state ownership, corruption, and poor economic management could strangle rapid growth, leading to greater poverty and a restive population.

CHAPTER TEN

Conclusions

This chapter highlights important insights from the preceding chapters. We begin with a summary of important domestic and near-abroad challenges for each of the three primary countries. We then describe common or related trends that are important to all of the nations. Finally, we focus on the likely implications for the U.S. defense establishment in general and the U.S. Navy in particular.

Major Future Domestic and Near-Abroad Challenges

The United States

The United States has today, and will have into the far term, by far the world's strongest economy. Unlike China, the United States will "get rich before it gets old." That does not mean that the United States will not experience significant challenges in 2015 and beyond. U.S. dependence on foreign energy sources shows no sign of changing. Tension resulting from the income disparity between the nation's rich and poor will also be an issue for future U.S. leaders. The most significant challenge for the United States will be its graying population.

Although the United States is and will remain a rich nation, the sheer scale of the needs of its aging population will put considerable strain on U.S. resources. As an increasing number of Americans enter the retired ranks, massive reallocations of government spending toward Social Security, Medicare, and Medicaid will take place (unless there is some major change in current U.S. entitlements policy). This reallocation of resources will constrain the nation's options in other areas, including defense spending. The Western European nations, with their extensive social-welfare structures and lower birthrates, started to face these resource allocation problems in the 1990s as the Cold War ended. It is noteworthy that in the absence of clear threats to their security, most European states sacrificed defense spending in favor of social programs to support their own aging populations. The United States will start to encounter similar problems after 2015. To an extent, the United States already has to confront these macro choices in national resource allocation, and defense is not winning the argument. Compared to the Cold War period, the United States today spends far less of its GDP on defense. That trend will likely intensify in the coming years as the country ages.

The United States has been fortunate since the Spanish-American War that it has not had to devote considerable military resources to defend its interests in the Western Hemisphere. It is likely that trend will continue, but the post-Castro endgame in Cuba, as well as the rise of anti-American leftist regimes in Central and northern South America, could, over time, result in a greater commitment of the U.S. military to the region. The greatest near-abroad security challenge to the United States will remain the inflow of illegal immigrants and illegal narcotics.

China

China is currently experiencing one of the most rapid economic expansions in modern times. Years of double-digit economic growth have benefited the Chinese in many ways. General standards of living have increased in most areas of the nation and the country is now the world's leading exporter. A favorable foreign-trade balance has allowed China to expand its domestic infrastructure and has resulted in a broader, more diverse economy than has existed at any time in China's past. The surge in economic power has also benefited the Chinese military, which now has far greater resources for recapitalization and modernization. Projections of China's 2020 GDP vary considerably, however, ranging from a low of $4 trillion to a high of more than $12 trillion.

Despite all the good news associated with its robust economy, the PRC still faces major, and growing, challenges. Of fundamental importance is the growing gap between the relatively rich coastal regions and the still-poor interior areas. This disparity has led to a huge migrant-worker class in China's eastern cities and to growing numbers of civil disturbances and protests, especially in the interior.

China's growing energy needs are of staggering proportions. As the nation modernizes, its energy needs will continue to grow in the coming decades. In the past three decades, most Asian nations have seen explosive growth in the number of automobiles on their roads. Today, a similar expansion is taking place in China, but on a far greater scale. The growth in auto usage, the rapid expansion of the number of factories, and the attempts to provide electricity to rural areas in the interior are all contributing to China's energy consumption. Currently, China is seeking to meet its energy needs through greater imports of oil and natural gas, mostly from the Middle East, and a dramatic increase in the number of domestic coal-fired power plants.

Pollution is a major challenge for China. As China's energy needs have expanded, the country has relied heavily on new coal-fired power plants. These new plants have contributed to a steadily growing air-pollution problem in and around the country's major urban areas. With levels of air and water pollution that would be considered totally unacceptable in most Western nations, China's pollution problems are likely to worsen in the coming two decades.

Finally, China faces the massive challenge of a graying population. Whereas most Western nations have a level of wealth that will, for the most part, allow them to cope with their own graying populations, China's situation is far more precarious. Most of the industrialized Western nations can say that "they got rich before they got old," but China "will get old before it gets rich." When this demographic problem is combined with China's energy and pollution issues, it is clear that China will face major domestic challenges by 2020–2025.

China's near abroad offers both potential and challenges for the nation. Mutual economic interests have allowed the PRC to improve relations with its southern and western neighbors. China's rapidly expanding energy needs will continue to draw it westward to an extent never seen in the country's history. The east and northeast, however, remain possible flash points as the ever-unpredictable North Korea continues to antagonize its neighbors and as the situation with Taiwan remains uncertain. Of major concern to China is whether U.S influence in East Asia wanes in the coming decades, and how Japan might react.

Iran

Iran's economy is performing well below its potential. Various societal and cultural factors contribute to this underperformance, but the important point is that without major governmental and public policy reforms, Iran will remain a relatively weak, single-sector (energy), export-dependent economy that will have major difficulty in meeting the expectations of its people. The fact that Iran also experiences a significant brain drain of educated professionals is both a symptom and cause of the nation's difficulties.

Energy revenues will help prop up the Iranian economy for the foreseeable future, but the lack of diversity in the economy will hinder overall growth. Fortunately for Iran, it will by 2020 have a more balanced population structure without the dramatic youth bulge that is so apparent in most other Muslim nations. This will help the nation, but the other economic challenges will still have to be overcome. In the near term, at least, there is little if any indication that the nation will take the steps necessary to reform its economy. Iran's imbalanced economic structure and the drain of skilled, educated professionals from the country will hinder Iran's ability to become the dominant power in the Middle East.

In its near abroad, Iran will continue to experience major challenges due to tensions with Israel, its Sunni neighbors, and the West, especially the United States. To the extent that Iran continues to antagonize its Sunni Arab neighbors, the United States, and the EU, it will remain relatively isolated from the West and the Arab Muslim world. At present, the "end game" in Iraq is still unclear. It may be possible for the Iranians to create a Shia ally state in central and southern Iraq, but even that is not certain. If tensions remain high in the Middle East, Iran will be forced to spend a considerable portion of its national income on defense, including the considerable effort required to build a nuclear arsenal. Perceived threats from its Sunni neighbors and the

United States will probably lead Iran to continue to seek a major ally in the form of China, Russia, or both.

Common Challenges

While each of the three nations has its own challenges, there are some common issues. How the three nations deal with these challenges could result in either greater tensions or increased opportunities for cooperation.

Aging

In the case of both China and the United States, graying populations will be a major challenge, particularly after 2020. China's One-Child policy and the relatively low birthrate of the United States, combined with the fact that people are living longer today, will result in grayer populations in both nations as the years pass. These grayer populations will require ever-greater social and health services that will require each nation to reallocate resources. How well each country's economy has grown—or not—between now and 2020–2025 will have a major impact on how much this reallocation of resources toward the older generation will affect other government spending in the United States and China. Of course, there is the possibility that military spending in both nations could be seriously affected, particularly if either nation does not perceive a clear threat to its security.

Iran will not be affected by the aging trend. Iran's challenge, so common to Muslim nations today, is the youth bulge and the associated problems of providing education and employment to large numbers of young people. In that regard, Iran's population profile after 2020 will start to look far more normal, with much of the current youth bulge gradually being eliminated.

Pollution

All three nations will continue to experience challenges from pollution as their populations continue to grow, urbanization increases, and their national energy needs expand. In the case of China, pollution challenges will be especially bad due to today's lax environmental standards and the rapid expansion of the use of coal-fired power plants. Pollution is already causing significant health problems in some industrial Chinese cities and even in some rural areas. Unless China makes major policy or technology changes, or both, this situation will worsen.

The United States and Iran will also continue to experience challenges in this area, as they do today, although their problems will be less severe than China's. Shared interest in the area of pollution could become a catalyst for mutual cooperation among the three primary nations. For example, the United States and the PRC both have huge domestic coal deposits. To the extent that both are concerned about the environment and pollution levels, there could be an opportunity for mutual research and technology-sharing to support "cleaner" use of their considerable coal reserves.

Energy

The energy needs of all three nations will continue to grow in the coming years. China's will grow fastest due to the nation's huge population and the rapid expansion of the nation's economy. Both China and the United States will have to increase their energy imports in the coming years, and most of that energy will come from the Middle East. In the case of Iran, the nation needs to diversify its economy and improve its domestic energy allocation and distribution system.

Coal is a possible area of Sino-U.S. cooperation in the energy sector, especially due to concerns over pollution. There is, however, the possibility that the growing competition for energy between the United States and China, much of it centered in the Middle East and the associated transit routes, could also lead to confrontation. The PRC's presence in Central Asia and the GME is expanding rapidly, mostly due to the country's growing energy needs. There is the possibility that in the coming years, something similar to the Cold War's competition in the region (where the Soviet Union vied with the West for influence among energy-rich nations) could occur again. Due to the size of its energy reserves and its strategic location at the entrance of the Straits of Hormuz, Iran will, of course, be an important player in the competition for energy.

Implications

What do these trends imply for the United States writ large and the Navy in particular? We first review likely strategic implications for the United States in general, then postulate their potential consequences for the Navy.

For the United States

It is a certainty that the average age of the U.S. population will continue to increase in the next two to three decades. Unless the U.S. economy experiences unexpectedly rapid growth prior to and during those years, a significantly greater portion of the nation's resources will have to be devoted to supporting the elderly population. The result could be pressure on the amount of funding available for defense. Given the high percentage of the population that will be in or approaching retirement age in 2020, it is likely that domestic political pressures will encourage considerable cutbacks in military spending.

It is interesting to note that many Western European nations have already gone through this process. Birthrates in many European countries have for decades been lower than that of the United States. This resulted in the aging of many of the European nations between 1980 and 1990. By the end of the Cold War in the early 1990s, there was considerable political pressure in much of Europe to reduce defense spending and transfer resources to social spending. Given their much more extensive social welfare programs (compared to those of the United States), and because their economies were not growing at the rate of the U.S. economy at the time, many European nations cut military spending dramatically. By the late 1990s, many of these nations were spending less than 2 percent of their GDPs on defense. In roughly two decades, the United States could find itself confronting similar hard choices.

Of course, if the aging nation perceives a serious external threat, it may still attempt to maintain high levels of defense spending. It appears unlikely, however, that the U.S. near abroad will be the source of a serious military threat to the nation.

U.S. dependence on imported energy will remain high and will probably continue to grow over the next two decades. A large percentage of the imports of oil and natural gas will continue to come from the Middle East. That guarantees significant U.S. strategic interests in that volatile region. To the extent that the GME is seen to be in turmoil, the United States might have to commit greater military resources compared even to today's levels.

The United States will continue to see challenges and opportunities in regions of interest. The rise of China will be the greatest strategic challenge facing the United States in the coming decades. The possible emergence of a Russian-Chinese de facto alliance would be of great concern to the United States. Events in the U.S. near abroad could also become more threatening if the rise of anti-American leftist regimes continues in the next two decades. Each of these strategic trends will have important implications for the U.S. Navy.

For the U.S. Navy

The Navy-specific implications of the strategic trends described in this monograph fall into six major categories. This section highlights what we consider to be the most likely and important implications for the Navy.

There will be less tolerance for costly, "big-ticket" defense projects. As pressures for social spending increase in the next two decades, resources for defense will come under greater scrutiny. Large programs (i.e., those that cost $10 billion or more in today's dollars) will probably be subjected to greater oversight and more demands from the public that the programs be truly necessary to the nation's defense.

China will remain the Navy's greatest potential challenge. The United States and the PRC are economic partners and competitors. It is still not clear whether the two nations will become serious military rivals. Nonetheless, both nations are clearly basing some of their military plans and modernization efforts on that premise. Of all the potential opponents of the United States, China has the greatest ability to build a modern, high-tech military that could be a serious challenge to the U.S. armed forces, including the Navy. Therefore, at the "high end" of the conflict spectrum, China is the most serious opponent for the Navy. As China's political, economic, and security interests continue to expand westward into the GME, the Navy could find itself confronted by hostile nations (such as Iran) that have been armed by China. It could even encounter a growing Chinese military presence in that area. In the far term, however, China's expanding military power will be most apparent in that nation's near abroad, including the surrounding seas.

Further cooperation with key allies will be required. The rise of China and the continued hostility of Iran will guarantee U.S. Navy involvement in the Pacific and GME regions. Therefore, the Navy should further its relationships and tackle military interoperability issues with key regional friends and allies. In the case of the Pacific, the relationship with the Japanese is fundamental to U.S. interests. The Navy should continue to pursue opportunities to improve its ability to interoperate with the Japanese Self Defense Force. In the case of the Middle East, the United States Navy may discover opportunities to further cooperation with key regional friends.

The Navy's "blue-green" mix will be affected. How the Navy adjusts its blue-green mix is related to how the United States perceives the threat from China. If China is seen as a rising military competitor, the Navy will probably have to continue to place most of its emphasis on its "blue-water" capabilities (i.e., traditional, high-intensity combat capabilities). In the particular case of a Chinese opponent, those capabilities will have to include the ability to fight and win against a nuclear-armed opponent. On the other hand, if other threats (such as low-intensity warfare in the U.S. near abroad or operations to counter Iranian-sponsored irregular warfare in the Middle East) are predicted to be the main mission in the next two to three decades, the Navy's ship mix will gradually be drawn in the "green-water" direction. For the present, the Navy should probably retain a mix of high- and low-end capabilities and orient a larger part of those capabilities toward the high end of the conflict spectrum.

An enduring commitment in the Middle East will be required. The fact that the United States will have to import an ever-greater amount of foreign oil and natural gas guarantees that it will have to maintain a considerable military presence in the GME. The Navy will be a vital part of that effort. If the political situation in the region makes it increasingly difficult to maintain sizable U.S. forces ashore in Muslim nations, more of the burden of basing those forces will be placed on the Navy.

Iran will continue to challenge the United States in the Middle East, probably in innovative ways. As of this writing, Iran is probably pursuing a nuclear-weapons program. It is certainly possible that Iran will be able to deploy an operational nuclear capability sometime before 2025. However, Iran's use of nuclear weapons would certainly be highly constrained by world opinion and the possibility of massive retaliation. At a minimum, U.S. options against a nuclear-armed Iran would be significantly more complicated compared to today. Simultaneously, Iran will probably continue to challenge the United States in the region through the use proxies such as Hezbollah and other radical, nonstate organizations. The use of nonstate proxies to undermine Western interests in the region will reduce the likelihood of direct U.S. action against Iran, but may also cause the region to remain in a state of turmoil for decades to come. U.S. Navy forces in the GME may experience terrorism, sabotage, protests from hostile groups in various countries, and other challenges as they conduct operations in that vital but volatile region.

Investment Implications for the U.S. Navy

As China's power expands in the next two decades, and as U.S., European, and Japanese dependence on Middle Eastern energy sources grows, the Navy will be expected to protect U.S. interests in both the Middle East and the Western Pacific. This means that the Navy will have to maintain sufficient numbers of ships—with capabilities adequate to deal with high-end crises in both regions—for the next two to three decades, if not longer. Shipbuilding budgets will have to support an adequate fleet size for these challenging, long-term missions.

If U.S. relations with China deteriorate over time, the Navy will have to focus considerable investments on the capabilities needed to defeat an increasingly high-tech adversary armed with considerable anti-access capabilities. This situation is the Navy's greatest potential challenge. The budget implications of having to be prepared to defeat an increasingly capable Chinese military are significant. Should the PRC elect to devote ever-greater resources to its military, the Navy will be required to field appropriate capabilities. For example, if the Chinese decide to field large numbers of sophisticated antiship missiles, the Navy would be forced to develop expensive countermeasures.

If the current emphasis on irregular warfare continues, more Navy investment in brown- and green-water capabilities will be needed. In most respects, this scenario is at the opposite end of the spectrum of a challenge from China. If the so-called Long War against radical Islam continues for the next two decades, the Navy will probably find itself investing more in littoral capabilities to support U.S. and coalition irregular warfare operations both ashore and in coastal areas. If the main planning paradigm is to be prepared to fight China, the Navy would probably have to significantly increase its missile-defense capabilities; in the opposite situation, the Navy would find itself investing more in sea basing and amphibious and coastal operations.

To the extent that regional powers, like Iran, acquire nuclear weapons, the Navy will have to devote resources to preparing for that eventuality. After the collapse of the Soviet Union, the Navy, like its sister services, devoted much less attention to the possibility of nuclear-weapons use. As states like North Korea and Iran have attempted to acquire this class of weapon, the possibility of U.S. forces having to prepare for combat against a nuclear-armed adversary has reemerged, although the number of nuclear weapons such countries could deploy is far smaller than the Soviet arsenal. To the extent that this threat continues to grow, the Navy will have to devote increasing resources to develop offensive and defensive capabilities. These capabilities, which include (1) hardened electronic systems capable of withstanding an electromagnetic pulse and (2) missile defense, are potentially expensive propositions.

Final Thoughts

This monograph has examined major domestic trends that are underway in the United States, China, and Iran in an attempt to determine how those trends might influence how each nation allocates resources at the national level. We also examined the near abroad of each nation (as well as the near abroads of Russia and Japan) to see whether there are significant changes under way that could affect the security situation of each nation in its immediate region.

Because the time horizon of this study was 20 or more years in the future, it is difficult to make precise predictions. Economies may grow more or less than current forecasts, unpredictable natural disasters or pandemics could consume considerable resources, and technological advances that are more rapid than expected in the energy industry might reduce reliance on imported fossil fuels. Any of these possibilities could change how the three countries allocate resources at the national level.

The overall age of the populations of both the United States and China will almost rise between now and 2025–2030. The increasing numbers of elderly citizens will have a major effect on how each country allocates fiscal and other resources and could result in a significant reprioritization of national budgets toward health care and other services for the elderly. Barring a clearly perceived security or other threat to either country, this profound demographic shift will put major pressure on defense spending in the far term.

The U.S. Navy's planning horizon for major new systems is a decade or more, and its air and ship platforms tend to be in service for 20–30 years after they are built. This means that the generation of platforms being designed and budgeted in 2015 and beyond will be influenced by the demographic trends described above. Chinese military planners will feel the effects of the resource shifts in roughly the same period.

Comparisons

This appendix compares the United States, Iran, and China in key domestic-trend areas. The United States and China are far larger than Iran in terms of their population, their economy, and other factors; their size has an effect on the challenges they face and the resources they can devote to problems. Iran is much smaller, with a much less-diversified economy than the United States or China. Nevertheless, this comparison provides a quick overview of the nature of the key domestic challenges facing each nation.

Table A.1
Comparison of Major Trends

Domestic Trend	United States	China	Iran
Population			
Growth rate	The United States will continue to experience better-than-replacement-level population growth due to birthrates and immigration.	China's population growth will place a greater burden on its domestic resources than will be the case in the United States.	Iran will experience a youth bulge through 2020 and a rapidly growing population of elderly residents through 2050.
Aging	The population of elderly U.S. residents will continue to grow, requiring the United States to increasingly divert resources toward social services.	By 2020, China will have more than 300 million people over the age of 65, placing considerable stress on existing social systems.	Iran's elderly population will triple after 2020.
Sex ratio	The United States will continue to maintain a normal male-to-female ratio.	China's current imbalance—young males outnumbering young females—will continue to worsen.	Iran's ongoing sex ratio favors males.
Youth population	The ratio of youth to other populations in the United States will remain normal.	China's youth cohort will be smaller than normal.	Iran's youthful population will represent between 20 percent and 23 percent of the total population through 2020.

Table A.1—Continued

Domestic Trend	United States	China	Iran
Environment			
Air quality	Air pollution levels will remain a concern in the United States, especially if the country increases its use of coal.	Air pollution is already a serious issue for many Chinese urban areas; the problem will worsen due to increasing reliance on coal.	Air pollution is Iran's most significant environmental problem; it is the result of Iran's energy sector, growing rate of urbanization, and young population.
Water	Selected problems in parts of the United States exist. In the far term (i.e., beyond 2020), rising water levels due to global warming could be a major issue.	There are serious water shortages in part of China. Water scarcity will become a more serious problem in the future.	Water availability has been decreasing in Iran, especially over the past several years. This has led to water shortages and rationing.
Land	The United States will experience coastal erosion and desertification.	China's land problems will be significantly greater than those of the United States. Desertification will be severe in some regions.	Arable land is decreasing in Iran because of changing rainfall patterns.
Economy			
GDP growth	The United States should continue to experience sustained GDP growth rates of 2–4 percent per year through the far term.	Estimates indicate that China's economy will continue to expand at a good rate through 2020 and beyond.	Most estimates suggest that Iran's GDP growth rate will remain stagnant; this will impede government efforts to create jobs.

Table A.1—Continued

Domestic Trend	United States	China	Iran
Per capita GDP	Real per capita GDP will continue to grow steadily, increasing the real wealth of most Americans.	Regional wealth distribution problems and the huge number of old citizens will constrain the growth of real income for many Chinese.	Real per capita GDP is expected to grow, but only modestly.
Employment	The employment outlook in the United States should remain good through the far term.	China will continue to experience challenges in creating adequate employment for its large population.	This is the most significant challenge facing Iran. The country needs to create hundreds of thousands of jobs per year to meet growing demand; it cannot meet these targets because of entrenched structural problems and international sanctions.
Economic diversification	Prospects remain excellent that the already well-diversified U.S. economy will remain strong.	Chinese economic diversification has been gradually improving over time, but it will not reach the same level of diversification and sophistication as the United States even in the far term.	Although Iran has a rentier economy, it is among the most-diversified rentier economies in the Middle East (not counting the smaller Gulf states).

Table A.1—Continued

Domestic Trend	United States	China	Iran
Energy			
Energy efficiency	Compared to Europe and Japan, U.S. output-to-energy ratios reveal relative inefficiency; but the United States is far more efficient than China.	Significant inefficiencies will persist through the near term. It is unclear whether China will be able to make major improvements through the far term.	Iran's oil sector is notable for its inefficiency.
Import dependency	The United States is highly dependent on energy imports and is likely to remain so through the medium and possibly far term. Its energy needs will increase and its domestic oil supplies will dwindle.	Dependency is a growing challenge for China, although the problem is less severe than in the United States.	Iran is a major energy producer, but government subsidies drive up consumer demand and therefore force Iran to rely on imports.
Human capital			
Brain drain	Brain drain is not a major issue for the United States. The American higher education system will likely remain the best in the world.	Brain drain is not a major problem for China, since little emigration takes place. The quality of the Chinese higher education system is an issue.	Brain drain is a major issue for Iran, which often has the highest rate of brain drain in the world due to poor employment prospects and government repression.

Table A.1—Continued

Domestic Trend	United States	China	Iran
R&D spending (as a percentage of GDP)	Although not as high as Japan, U.S. rates of R&D investment remain high. The quality of U.S. higher education contributes to this high level of investment.	Spending is growing but remains patchy in particular areas of the economy.	Spending is minimal in all sectors, especially the energy sector.
Primary education	Primary education is generally good in the United States, although the United States may continue to lag behind Europe and Japan. U.S. primary education is far superior to that offered in China.	Primary education is a significant issue in China; quality varies greatly by region. Near- and medium-term prospects are unclear.	Iran has an excellent primary education system.
Other			
Ability to respond to shocks	The overall ability of the U.S. economy to respond to shocks is excellent.	China is and will remain vulnerable to shocks (such as export disruptions).	Iran is vulnerable to external shocks because of its rentier economy; to date, it has managed those shocks.

China's Coal Future

As observed in the main body of this monograph, China depends on coal for about 70 percent of its energy and has the world's third-largest coal reserve (behind the United States and Russia). China is the world's largest producer of coal, and until recently was a net exporter of coal. Demand for coal in China has increased with China's economic growth and has outstripped supply. For example, Shanxi province, which produces a quarter of China's coal, recently experienced a 15-percent coal shortage for electrical-power generation. Although coal companies in Shanxi have increased coal production by 9 percent, supply has not met demand.[1] Nationwide, China's coal reserves are shrinking, and rising coal prices are fueling inflation and cutting profits.[2] Recent independent analyses suggest that if China's economy continues to grow, China's coal production will peak between 2015 and 2025 and that production will significantly decline (to below current levels) by 2030–2040.

China's Coal Industry

Because China has always been largely self-sufficient in coal, major forces for supply and demand have been internal. The history of China's coal industry begins with organized production in 475 BC. China's

---

[1] "Supply Crunch Hits Coal Cradle," *China Daily*, July 9, 2008.

[2] "Power Coal Reserve Falls to 12 Days Amid Rising Prices," *China Daily*, April 23, 2008.

annual coal production increased from 40 million metric tons (tonnes) in 1950 to nearly 500 million tonnes in 1975. Under a Stalinist-Maoist doctrine of quantity above all else, efficiency in coal extraction was not a consideration, and much of the extracted coal was wasted. Output from large mines was emphasized, with more than 400 new mines started in 1958. In that year, China more than doubled coal production, operating approximately 110,000 pits.[3] Over a long and sometimes profligate history of coal production, China's largest and most accessible sources of coal have been mined out.

China's coal production was reduced between 1997 and 2000 as China shut down about 46,000 small (and often illegal) coal mines. Shifting government policies and GDP growth since 2000 have been cyclic, shaping forces on China's coal industry. As of 2004, China had 28,000 coal mines, 95 percent of which were small (i.e., producing less than 30,000 tonnes per year) and only 40 percent were mechanized. Manual coal production techniques in most Chinese mines remain inefficient, recovering only about 30 percent of the mined coal. In 2006, China's mines extracted on average 400 tonnes of coal per worker—an amount equal only to about 5 percent of levels achieved in coal-exporting countries such as Australia. Some coal-extraction problems in China stem from shortages of extraction and loading equipment. Coal production has been further hampered by limited supporting transportation infrastructure. Recent pressure for additional coal has also tended to reduce the efficiency of coal extraction in China. In 2003 and 2004, most Chinese coal consumption was for power, and approximately 70 percent of China's power was generated by thermal coal. Metallurgy, cement, export, and other uses (such as home heating) accounted for smaller percentages of coal usage. China has been the world's primary coke producer (it accounted for 46 percent of world coke production in 2003), exporter, and consumer.[4] Recently, however, China has experienced increasing domestic demand for coke. China has therefore decreased coke

---

[3] Vaclav Smil, *China's Past, China's Future: Energy, Food, Environment*, New York: Routledge Curzon, 2004.

[4] Coke is the best source of the carbon needed to convert iron into carbon steel.

resources and production and has begun cutting back on coke exports by increasing coke export tariffs.[5]

A Simple Model of China's Future Coal Production

China's coal reserve of proved recoverable coal is variously estimated at between 110 billion and 125 billion tonnes.[6] For the sake of analysis, we assume a 125 billion tonne reserve. China has reported that 2.38 billion tonnes of coal were produced in 2006. At that rate of production, China's coal reserves will be exhausted in 53 years, or around 2060. China will, of course, conduct exploration to find new coal reserves. New reserves, however, could be expensive to extract and be of low quality (i.e., low energy content). These factors—rate of depletion, finding new reserves, and their quality—will greatly influence how much coal China will have to import in the coming decades.

The difficulties of predicting China's coal future are apparent from our own simple analysis. Perhaps the most obvious difficulty is uncertainty about the amount of China's coal reserves. Given the 10-percent discrepancy between high and low estimates about these reserves, it is extremely difficult to estimate China's future coal-production rate. The most certain thing about the problem is that coal production will not remain at the 2006 level.

In China's race to become as rich as possible before it gets old, Chinese coal production is growing rapidly, averaging 10.3-percent annual growth since 2001 (see Figure B.1).[7] China's National Coal Association has predicted that the country will need about 2.5 billion

---

[5] Guanghua Liu, *China's Coal Supply/Demand and Their Impact on International Coal*, AAA Minerals International, undated.

[6] Proved recoverable reserves are the tonnage of coal that can be recovered under present and expected local economic conditions with existing available technology.

[7] The 1996–1999 "notch" in this figure reflects the closure of small Chinese coal mines starting in 1997.

Figure B.1
China's Coal Production Since 1970

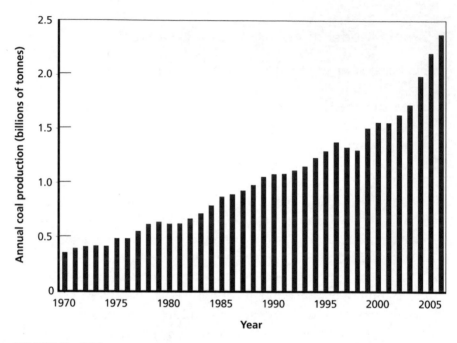

SOURCE: Tu, 2006.
RAND *MG729-B.1*

tonnes of coal in 2010. As a hedge, China plans to increase coal production *capacity* to 3 billion tonnes per year by then.[8]

To explore the role of growth rate in coal production, we assume
that coal production will grow at constant rates *until current estimated
reserves are exhausted*. The long-term implications of annual growth
rates between 0 percent and 5 percent are shown in Figure B.2. The
"no-growth" line represents the result of production remaining at the
2006 level; zero growth results in the exhaustion of coal reserves in
2060. The very modest annual growth rate of 1 percent significantly
accelerates the date of coal reserve exhaustion, advancing it by about
ten years. This is an example of "the miracle of compound growth."

---

[8]   "China's Annual Coal Production Capacity to Exceed 3.1 Bln Tons by 2010," *China
Daily*, June 15, 2007.

Figure B.2
Projections of China's Cumulative Coal Production from 2005

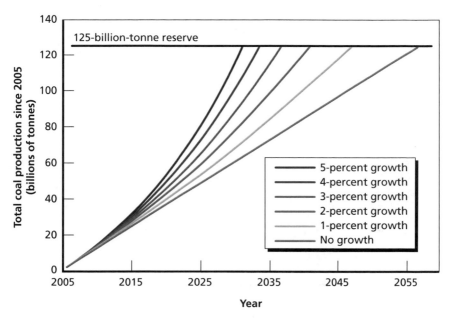

RAND *MG729-B.2*

This ten-year reduction in the total time to coal reserve exhaustion equals a nearly 20-percent change, a total far greater than the roughly 10-percent uncertainty in coal reserves. An annual growth rate of 5 percent halves the time to reserve exhaustion (to about 26 years). This too suggests that more attention should be paid to production rates than to coal-reserve estimates.

These initial estimates ignore the reality that at a certain point, coal production becomes progressively more difficult. Production levels for resources such as coal or oil have been observed to follow bell-shaped curves (see Figure B.3). Roughly speaking, peak production occurs when reserves have been depleted by 50 percent—a far more realistic assumption than constant growth until reserves are exhausted.

Bearing this mind, we focus on near-term growth in coal production rather than growth to coal exhaustion. Here we note that a near-term growth rate in coal production of about 10 percent is needed to accommodate China's planned economic growth. Also, as noted

Figure B.3
Historical British Coal Production

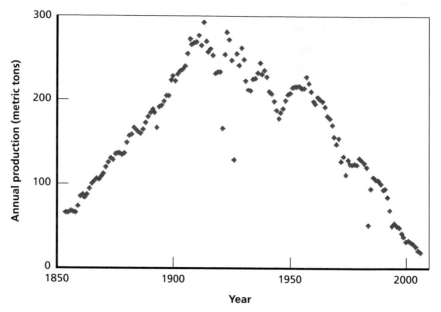

SOURCE: Data from the U.S. National Bureau of Economic Research (1854–1876), the Durham Coal Mining Museum (1877–1956), and the British Department of Trade and Industry (1957–2006).

RAND MG729-B.3

above, China's coal production has grown at an average annual rate of 10.3 percent since 2001 (the year after elimination of small mines was completed). With a 10-percent near-term annual growth rate in coal production, half of China's current 125-billion-tonne coal reserve will be consumed by 2018 (see Figure B.4). Note also that with an 8-percent near-term annual growth rate in coal production, half of China's coal reserve will be consumed just one year later (i.e., 2019). Simple analysis then suggests that China's coal production might peak in 2018.[9]

---

[9] Notice that by focusing on the 50-percent depletion level, our methodology avoids the problem of increasingly difficult coal production as coal reserves approach total exhaustion.

Figure B.4
Projections of Time to Peak Coal Production in China

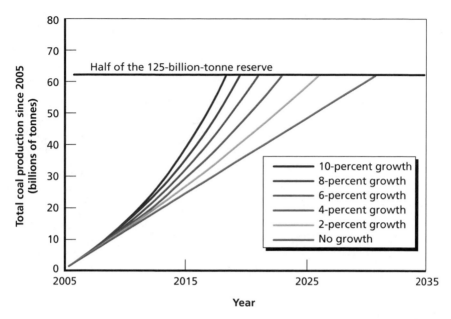

International Energy Outlook 2007—Energy Information Administration

The 2007 international energy outlook report by the EIA is relatively optimistic about the future of China's coal.[10] The EIA assumes that China's coal needs will be met by domestic production through 2030.[11] The report predicts a 2.9-percent growth rate in China's coal production between 2004 and 2030.[12] (Note, however, that a growth rate of 2.9 percent will be inadequate to sustain China's expected economic

---

[10] Energy Information Administration, *International Energy Outlook, 2007*, May 2007a.

[11] Energy Information Administration, 2007a, p. 50.

[12] The EIA report generally works in British thermal units (Btu), which do not translate directly to tonnes of coal because differing types of coal have differing energy contents. As a result, detailed results from this report are not presented here.

growth through 2030.) Accepting the growth rate of 2.9 percent a year through 2030, the above methodology indicates that China's coal production will peak in 2025. The EIA's estimate of China's coal reserve, 126.2 billion tons (i.e., 114.5 billion tonnes) in 2003, is consistent with other assessments of China's coal reserve. The report adds that

> [w]hereas China in the past had offered an export tax rebate of 8 percent to encourage exports, it has now imposed a 5-percent export tax on coking coal and may apply an export tax on steam coal in the future. China has also lowered its export cap to 46 million tons (1.1 quadrillion Btu) for 2007, equivalent to about one-half of China's steam coal exports in 2003. Australia, Indonesia, and other suppliers are projected to compensate for the shortfalls in China's coal exports, as occurred in 2005 when China reduced its exports by 16 million tons (0.4 quadrillion Btu) from their 2004 level.[13]

World Energy Outlook 2006—International Energy Agency

The IEA's 2006 World Energy Outlook (WEO) stresses that the overwhelming share of global increases in coal use by 2030 will come from China.[14] At the same time, it assumes that China will remain a net exporter of coal throughout the next 25 years, but that China will lose market share as more of its output is diverted to its rapidly growing domestic market. According to the WEO, Chinese demand will rise from 1.881 million tonnes per year in 2004 to 2.603 million tonnes in 2010, 3.006 million tonnes in 2015, and 3.867 million tonnes in 2030. The estimated average annual growth rate of consumption between 2004 and 2030 is 2.8 percent. The WEO also projects domestic production to increase at an average annual growth rate of 2.7 percent, or from 1.960 million tonnes in 2004 to 2.673 million tonnes in 2010, 3.074 million tonnes in 2015, and 3.927 million tonnes in 2030.

---

[13] Energy Information Administration, 2007a, p. 58.

[14] International Energy Agency, 2006.

These figures are consistent with the report's conclusion that China will remain a net exporter of coal until 2030. Regarding coal reserves, the WEO uses British Petroleum (BP) figures and estimates Chinese reserves at 118 billion tonnes in 2005, which is in line with figures used in other studies.

Assuming the WEO's annual growth rate of 2.7 percent and a reserve base of 125 billion tonnes, China's production will peak in 2028 at the latest. Compared to other recent reports, the WEO's projections appear conservative. In suggesting a relatively late peak, the WEO is optimistic about China's ability to remain a net exporter of coal.

The Future of Coal—The DG JRC Institute for Energy

The DG JRC Institute of Technology released a report on the worldwide future of coal to the European Commission in February 2007.[15] This report contains several findings relevant to this monograph. It breaks down countries and regions into four tiers, depending on how each area will be affected by the availability of coal imports between 2030 and 2070. Tier-1 areas (e.g., the United States and Russia) are self-sufficient or better. Tier-2 areas are prime exporters (i.e., Australia, Colombia, and Indonesia). Tier-3 areas are able to meet domestic demand (e.g., India). Tier-4 areas (e.g., China and many European nations) are facing potential shortfalls. The report states that:

> China has reserves that are adequate only to about 2030 and although significant coal deposits exist, considerable exploration and mining at depth (with associated major capital investment) will have to be undertaken this then reflects on the financial return.

> We calculate that China has 30 years of life in hard coal reserves and it is difficult to see if production can be maintained at over 2 Btpa [billions of tonnes per annum] past 2030 or 2040, which

---

[15] B. Kavalov and S. D. Peteves, *The Future of Coal*, DG JRC Institute for Energy, 2007.

will focus attention on technological change, including nuclear and use of low quality, expensive-to-mine coal deposits.[16]

Referring back to Figure B.1, this implies that by 2030 or 2040, China's coal production will drop below its 2005 level. The report further states that it sees "the domestic markets of US and Russia enjoying a plentiful coal supply to 2050 but increasing pressure growing in India and to a much larger extent in China, especially post 2030."

Coal Resources and Future Production—The Energy Watch Group

Coal: Resources and Future Production was produced in 2007 by the Energy Watch Group (EWG). As a global study, one of the report's most interesting findings is the extent to which China's coal consumption rate is out of proportion to its recoverable reserves (see Figure B.5).

For comparison, note that China's annual production rate of 1.9 percent of reserves in 2005 is about three times the global average. Following country-by-country projections of world coal production, the EWG produced the projection shown in Figure B.6. Projections out to 2030 are bounded in this graphic by the WEO 2006 reference scenario, which continues current practices, and an alternative scenario in which demand and production are scaled back by about 10 percent. This figure contains the production estimate for China, which is divided into Chinese production of bituminous, subbituminous, and lignite coal. China's coal production is seen to peak relatively early in this figure because of China's high ratio of production to reserves. China's coal production is separated out in Figure B.7, which groups bituminous and subbituminous coal production; lignite coal production is shown separately.[17] Increased dependence on low–energy

---

[16] Kavalov and Peteves, 2007, pp. 49, 56.

[17] There are four major types of coal. Lignite is the softest coal, with the highest water content and lowest energy density (about 13 million Btu/ton). Subbituminous coal has lower water content and greater energy density (17 million to 18 million Btu/ton) than lignite. Bituminous coal has still lower water content and greater energy density (about 24 million

Figure B.5
Global Coal Reserves and Production Rates

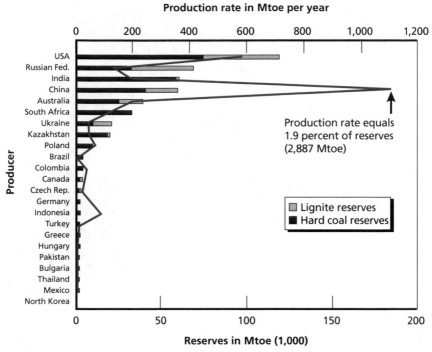

SOURCE: Energy Watch Group, *Coal: Resources and Future Production*, EWG-Series No. 1/2007, March 2007.
RAND MG729-B.5

density lignite, as shown in Figure B.7, suggests further pressure on China's domestic coal supplies and the need for increased imports to meet energy demands.

The EWG report states that it is "likely that China will experience peak production within the next 5–15 years, followed by a steep decline." As shown in Figure B.7, this equates to a peak in production between 2012 and 2022.

Btu/ton). Anthracite has the lowest water content and the highest energy density (about 25 million Btu/ton). China does not produce meaningful amounts of anthracite.

Figure B.6
Possible Worldwide Coal Production

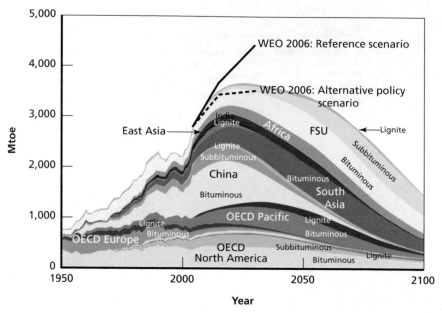

SOURCE: Energy Watch Group, 2007.
RAND *MG729-B.6*

British Petroleum Statistical Review of World Energy, 2007

BP produces an extensive annual statistical review of world energy reserves and consumption. In terms of demand, the current review finds that in 2004, China had an estimated 271 gigawatts of coal-fired capacity in operation. To meet the demand for the electricity that is required to sustain rapid economic growth, China will bring an additional 497 gigawatts of coal-fired capacity by 2030. This will require large financial investments in new coal-fired power plants and associated transmission and distribution systems. This study also describes an emerging source of demand for coal in China.

With a substantial portion of the increase in China's demand for both liquids and natural gas projected to be met by imports, the

Figure B.7
Coal Production in China Based on Current Reserve Estimates

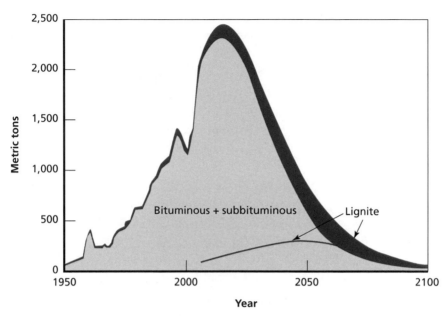

SOURCE: Energy Watch Group, 2007.
RAND *MG729-B.7*

Chinese government is actively promoting the development of a large coal-to-liquids industry. Initial production of coal-based synthetic liquids in China began in late 2007 with the completion of the country's first coal-to-liquids plant (located in the Inner Mongolia Autonomous Region). The plant was built by the Shenhua Coal Liquefaction Corporation and has an initial capacity of approximately 20,000 barrels per day, tentatively scheduled to be increased to 100,000 barrels per day by 2010.

The 2007 edition of the BP report states that China has a coal reserve of 114.5 billion tonnes, with 48 years until reserve exhaustion at the current rate. In the context of our simple analysis, this suggests a peak in coal production between 2015 and 2020.

Conclusions

This appendix characterized China's coal industry, presented RAND's simple analysis of China's coal future, and described credible, current studies of China's coal future.

After nearly 2,500 years of coal production, China's largest and most accessible sources of coal have been mined out. China's coal industry depends on small and inefficient mines whose coal is usually extracted manually. There is no reason to believe that China will find large new coal deposits or that Chinese coal production will become significantly more efficient.

Our simple analysis of China's coal future assumes a 10-percent near-term growth rate consistent with growth since 2000 and indicates that peak production should occur in approximately 2018.

A Chinese forecast of China's coal future, stripped of its disingenuous start date, suggests that China's current estimated coal reserves will be depleted between 2047 and 2052. China recently limited coal exports and changed its tax rates for coal export (to discourage coal exports). These changes occurred as China became—for the first time in its history—a net importer of coal.

Collectively, these facts suggest that China's coal production will peak between 2015 and 2025, with a significant decline (to below current levels) by 2030–2040.

Bibliography

Abbasi-Shavazi, Mohammad Jalal, "Recent Changes and the Future of Fertility in Iran," in *Completing the Fertility Transition*, New York: United Nations, 2002.

Abbaspour, M., and A. Sabetraftar, "Review of Cycles and Indices of Drought and Their Effect on Water Resources, Ecological, Biological, Agricultural, Social and Economical Issues in Iran," *International Journal of Environmental Studies*, Vol. 62, No. 6, December 2005.

Alaedini, Pooya, and Mohamad Reza Razavi, "Women's Participation and Employment in Iran: A Critical Examination," *Critique: Critical Middle Eastern Studies*, Vol. 14, No. 1, Spring 2005.

Alamdari, Kazem, "The Power Structure of the Islamic Republic of Iran: Transition from Populism to Clientelism, and Militarization of the Government," *Third World Quarterly*, Vol. 26, No. 8, 2005.

Alizadeh, Parvin, "Iran's Quandary: Economic Reforms and the 'Structural Trap,'" *The Brown Journal of World Affairs* Vol. 9, No. 2, Winter/Spring 2003.

Allisa, Sufyan, "The Challenge of Economic Reform in the Arab World: Toward More Productive Economies," Carnegie Endowment for International Peace, *Carnegie Papers*, No. 1, May 2007.

Amuzegar, Jahangir, "Iran's Crumbling Revolution," *Foreign Affairs,* Vol. 82, No. 1, January/February 2003.

Arbitrio, Roberto, et al., "I.R. of Iran 2006–2008," *Strategic Programme Framework,* United Nations Office on Drugs and Crime, October 2005.

Asadollah-Fardi, G., "Air Quality Management in Tehran." As of June 19, 2008: http://www.iges.or.jp/kitakyushu/mtgs/seminars/theme/uaqm/Presentations/Tehran/Tehran.pdf

Aslaksen, Silje, and Ragnar Torvik, "A Theory of Civil Conflict and Democracy in Rentier States," *Scandinavian Journal of Economics,* Vol. 108, No. 4, 2006.

Bahgat, Gawdat, "Nuclear Proliferation: The Islamic Republic of Iran," *Iranian Studies,* Vol. 39, No. 3, September 2006.

Bank of Finland, *BOFIT Weekly*, Vol. 28, July 12, 2007.

Bea, Keith, *Federal Stafford Act Disaster Assistance: Presidential Declarations, Eligible Activities and Funding*, Congressional Research Service, Order Code RL33053, August 2005.

Beblawi, Hazem, "The Rentier State in the Arab World," in Hazem Beblawi and Giacomo Luciani, eds., *The Rentier State*, New York: Croom Helm, 1987.

———, "The Rentier State in the Arab World," in Giacomo Luciani, ed., *The Arab State*, Berkeley: University of California Press, 1990.

Bergheim, Stephan, "Global Growth Centres 2020: Challenges and Choice for European Policymakers," *Deutsche Bank Research*, November 2006.

Bergsten, C. Fred, Bates Gill, Nicholas R. Lardy, and Derek J. Mitchell, *China: The Balance Sheet*, New York: CSIS & Institute for International Economics, Public Affairs, 2006.

Bozorgmehr, Majmeh, "Picture Diary: 5 Days with Ahmadi-Nejad," April 22, 2007. As of June 16, 2008:
http://www.ft.com/cms/s/0/5ed75b10-f0f9-11db-838b-000b5df10621.html?nclick_check=1

British Petroleum, *Statistical Review of World Energy*, London, 2006a.

———, *World Statistical Yearbook 2006*, London, 2006b.

Bureau of Economic Analysis, "National Economic Accounts," last updated August 28, 2008. As of September 12, 2008:
http://www.bea.gov/national/index.htm#gdp

Center on Budget and Policy Priorities, *Future Medicaid Growth Is Not Due to Flaws in the Program's Design, But to Demographic Trends and General Increases in Health Care Costs*, white paper, February 2005.

Central Intelligence Agency, "Iran," *World Factbook*, 2003–2007.

"China Bank Bailout Could Need US$290 Bln: Report," *People's Daily Online*, January 27, 2003. As of June 2, 2008:
http://english.people.com.cn/200301/27/print20030127_110839.html

"China Detains Four HIV-Positive People Asking for Help," Agence France-Presse, July 15, 2004.

"China Fears Brain Drain as Its Overseas Students Stay Away," *The Guardian* (London), June 5, 2007.

"China: The Shifting Strategy on Korea," Stratfor, April 23, 2007.

"China: Yangtze Is Irreversibly Polluted," Associated Press, April 15, 2007.

"China's Annual Coal Production Capacity to Exceed 3.1 Bln Tons by 2010," *China Daily*, June 15, 2007.

"China's Health Care: Where are the Patients?" *The Economist*, August 19, 2004.

Chinese Academy of Social Sciences, *Blue Book of Chinese Society: 2004*, Beijing: Social Sciences Documentation Publishing House, 2004.

Cody, Edward, "Hu Set for Second Term at China's Helm," *The Washington Post*, October 14, 2007.

Congressional Budget Office, *A 125-Year Picture of the Federal Budget's Share of the Economy*, Long-Range Fiscal Policy Brief, revised July 2002a.

———, *The Looming Budgetary Impact of Society's Aging*, Long-Range Fiscal Policy Brief, July 2002b.

———, *The Long-Range Budget Outlook*, December 2003.

———, *The Economic and Budget Outlook: Fiscal Years 2006–2015*, Washington, D.C.: Congressional Budget Office, 2005.

———, *The Long-Term Implications of Current Defense Plans: Detailed Update for Fiscal Year 2007*, April 2007.

CountryWatch, "Iran: 2007 Country Review," 2007.

Crane, Keith, Roger Cliff, Evan Medeiros, James Mulvenon, and William Overholt, "Modernizing China's Military: Economic Opportunities and Constraints," unpublished RAND Corporation research, 2005.

Daggett, S., *Defense Budget: Long-Term Challenges for FY2006 and Beyond*, Congressional Research Service, Order Code RL32877, April 20, 2005.

Davoudpour, Hamid, and Mohammad Sadegh Ahadi, "The Potential for Greenhouse Gases Mitigation in Household Sector of Iran: Cases of Price Reform/Efficiency Improvement and Scenario for 2000–2010," *Energy Policy*, Vol. 34, 2006.

Doyle, Timothy, and Adam Simpson, "Traversing More Than Speed Bumps: Green Politics Under Authoritarian Regimes in Burma and Iran," *Environmental Politics*, Vol. 15, No. 5, November 2006.

Drew, Jill, "Protests in China Target French Stores, Embassy," *The Washington Post*, April 20, 2008.

Dueck, Colin, and Ray Takeyh, "Iran's Nuclear Challenge," *Political Science Quarterly*, Vol. 122, No. 2, Summer 2007.

Economist Intelligence Unit, "Country Report Iran," 2006.

———, "Market Indicators and Forecasts," 2007a.

———, "Country Report: China," June 2007b.

———, "China Economy: Critical Issues—Export Risks," June 22, 2007c.

Economy, Elizabeth, *The River Runs Black: The Environmental Challenge to China's Future,* Ithaca, N.Y.: Cornell University Press, 2004.

————, "Environmental Enforcement in China," in Kristen A. Day, ed., *China's Environment and the Challenge of Sustainable Development,* Armonk, N.Y.: M.E. Sharpe, 2005.

Energy Information Administration, *Country Analysis Brief: China,* August 2006a. As of May 20, 2008:
http://www.eia.doe.gov/emeu/cabs/China/Oil.html

————, *Country Analysis Briefs: Iran,* August 2006b. As of June 3, 2008:
http://www.eia.doe.gov/emeu/cabs/Iran/pdf.pdf

————, *International Energy Outlook, 2007,* May 2007a. As of June 16, 2008:
http://www.eia.doe.gov/oiaf/ieo/index.html

————, *International Energy Annual 2005,* June–October 2007b. As of May 20, 2008:
http://www.eia.doe.gov/iea/

Energy Watch Group, *Coal: Resources and Future Production,* EWG-Series No. 1/2007, March 2007.

England, Robert Stowe, *Aging China: The Demographic Challenge to China's Economic Prospects,* Westport, Conn.: Praeger, 2005.

Esfandiari, Golnaz, "Iran: Coping with the World's Highest Rate of Brain Drain," Radio Free Europe/Radio Liberty, March 8, 2004.

Eurasia Group 2006, "China's Overseas Investment in Oil and Gas Production," prepared for the U.S.-China Security and Economic Review Commission, October 16, 2006.

European Commission (DG Trade), *EU Bilateral Trade and Trade with the World,* 2007. As of June 16, 2008:
http://trade.ec.europa.eu/doclib/docs/2006/september/tradoc_113440.pdf

European Commission, *Russian Federation—Country Strategy Paper 2007–2013,* undated. As of June 16, 2008:
http://ec.europa.eu/external_relations/russia/csp/2007-2013_en.pdf

Fan, Maureen, "Illiteracy Jumps in China, Despite 50-Year Campaign to Eradicate It," *The Washington Post,* April 27, 2007.

Farrell, Diana, and Andrew J. Grant, "China's Looming Talent Shortage," *The McKinsey Quarterly,* No. 4, 2005.

Fetini, Habib, et al., "Iran: Medium Term Framework for Transition," Social and Economic Development Group, Middle East and North Africa Region, Washington, D.C.: World Bank, April 30, 2003.

Fewsmith, Joseph, "China and the Politics of SARS," *Current History*, Vol. 101, No. 656, September 2003.

Ford, Peter, "Tai Lake Algae Bloom Cuts Off Water to Millions in China's Jiangsu Province," *Christian Science Monitor*, June 4, 2007.

Freedom House, "Iran," *Freedom in the World*, 2006. As of June 16, 2008: http://www.freedomhouse.org/template.cfm?page=22&year=2006&country=6982

————, "Iran," *Freedom of the Press*, undated. As of June 16, 2008: http://www.freedomhouse.org/template.cfm?page=251&year=2006

Fuller, Graham E., "The Youth Factor: The New Demographics of the Middle East and the Implications for U.S. Policy," The Brookings Project on U.S. Policy Towards the Islamic World, The Saban Center for Middle East Policy, No. 3, June 2003.

Furman, Jason, *Options to Close the Long-Run Fiscal Gap*, testimony before the United States Senate Committee on the Budget, January 31, 2007.

Gazprom, *Annual Financial Report 2006*, Moscow, 2007.

"Geopolitical Diary: U.S., Iran Lose from Major Changes in Baghdad," Stratfor, April 17, 2007.

Gheissari, Ali, and Vali Nasr, "The Conservative Consolidation in Iran," *Survival*, Vol. 47, No. 2, 2005.

Global Investment House, "Iran Economic & Strategic Outlook: Well Diversified Economic Base," June 2007.

Goldstein, Morris, and Nicholas Lardy, "A Modest Proposal for China's Renminbi," *Financial Times* (London), August 26, 2003.

Goodman, Peter S., "Too Fast in China? Stunning Growth May Have a Built-In Problem: Overcapacity," *The Washington Post*, January 26, 2006.

Gries, Peter Hays, *China's New Nationalism: Pride, Politics, and Diplomacy*, Los Angeles: University of California Press, 2004.

"Group Says China Increased Arrests, Violence Against HIV Positive Protestors," Agence France-Presse, July 9, 2003.

"Guangzhou Threatened by Labor Shortage," *China Daily*, September 4, 2007.

Harrison, Frances, "Quake Experts Urge Tehran Move," BBC News, March 14, 2005. As of June 3, 2008:
 http://news.bbc.co.uk/2/hi/middle_east/4346945.stm

Herb, Michael, "No Representation Without Taxation? Rents, Development and Democracy," paper, Georgia State University, December 3, 2003. As of June 3, 2008: http://www2.gsu.edu/~polmfh/herb_rentier_state.pdf

The Heritage Foundation, *Federal Revenue and Spending: A Book of Charts*, undated. As of May 30, 2008:
http://www.heritage.org/research/features/BufgetCahrtBook/index.html

"The Hitch in Japan's Normalization Plan," Stratfor, September 12, 2007.

"How Fit Is the Panda?" *The Economist*, September 29, 2007.

Hudson, Valerie M., and Andrea M. den Boer, *Bare Branches: Security Implications of Asia's Surplus Male Population*, Cambridge, Mass.: MIT Press, 2004.

Illariounov, Andrei, "Russia Inc.," *New York Times,* February 4, 2006.

International Energy Agency, *World Energy Outlook 2004,* Paris, 2004.

———, *World Energy Outlook 2006*, Paris, 2006.

International Monetary Fund, World Economic Outlook Database, 2006a.

———, *Russian Federation: 2006 Article IV Consultation—Staff Report; Staff Statement; and Public Information Notice on the Executive Board Discussion,* Country Report No. 06/429, Washington, D.C., December 2006b.

———, *Islamic Republic of Iran: 2006 Article IV Consultation—Staff Report; Staff Statement; Public Information Notice on the Executive Board Discussion; and Statement by the Executive Director for the Islamic Republic of Iran,* IMF Country Report No. 07/100, Washington, D.C., March 2007.

Iranian Studies Group at MIT, "Earthquake Management in Iran, a Compilation of Literature on Earthquake Management, Draft," January 6, 2004. As of March 25, 2008:
http://www.vojoudi.com/earthquake/management/management_eq_mit_eng.htm

"Iran's Decision to Raise Gas Prices Exposes Economic Vulnerability," Associated Press, May 24, 2007.

Ivanov, Vladimir, *Russian Energy Strategy 2020: Balancing Europe with the Asia-Pacific Region*, ERINA Report, Vol. 53, 2003.

Jane's Information Group, "Going Soft, China's Alternative Route to Regional Influence," *Jane's Intelligence Review*, July 1, 2007.

Japan Ministry of Internal Affairs and Communications, *Statistical Handbook of Japan*, 2006, Chapter Two, "Population."

Jbili, Abdelali, Vitali Kramarenko, and José Bailén, *Islamic Republic of Iran: Managing the Transition to a Market Economy*, Washington, D.C.: International Monetary Fund, 2007.

Kalish, Ira, *Global Economic Outlook 2007*, Cambridge, Mass.: Deloitte Research, 2007.

Kavalov, B., and S. D. Peteves, *The Future of Coal*, DG JRC Institute for Energy, 2007. As of June 23, 2008:

http://ie.jrc.ec.europa.eu/publications/scientific_publications/2007/EUR22744EN.pdf

Kekic, Laza, "The *Economist* Intelligence Unit's Index of Democracy," 2007. As of June 16, 2008:
http://www.economist.com/media/pdf/DEMOCRACY_INDEX_2007_v3.pdf

Khosrokhavar, Farhad, "The New Intellectuals in Iran," *Social Compass,* Vol. 51, No. 2, 2004.

Langton, Christopher, *The Military Balance 2007*, The International Institute for Strategic Studies, London: Routledge, 2007.

Lardy, Nicholas, "China: Rebalancing Economic Growth," Peterson Institute for International Economics, 2007.

Larsen, Janet, "Iran's Birth Rate Plummeting at Record Pace: Success Provides a Model for Other Developing Countries," Earth Policy Institute, December 28, 2001. As of June 3, 2008:
http://www.mnforsustain.org/iran_model_of_reducing_fertility.htm

Lewis, Barbara, "China About to Become Biggest CO_2 Emitter: IEA," Reuters, April 18, 2007.

Liang, Dong, "Relationship Between Officials and the Masses," in *Blue Book of Chinese Society: 2004*, Chinese Academy of Social Sciences, Beijing: Social Sciences Documentation Publishing House, 2004.

Liang, Zai, and Zhongdong Ma, "China's Floating Population: New Evidence from the 2000 Census," *Population and Development Review,* Vol. 30, No. 3, September 2004.

Li, Rongxia, "The Medical Reform Controversy," *Beijing Review,* Vol. 48, No. 38, September 2005.

Liu, Guanghua, *China's Coal Supply/Demand and Their Impact on International Coal*, AAA Minerals International, undated. As of June 22, 2008:
http://www.aaamineral.com/Coal/ppt/Aruba%20George.ppt

Liu, Yuanli, "Development of the Rural Health Insurance System in China," *Health Policy and Planning,* Vol. 19, No. 3, May 2004.

Li, Zi, "Medical Reform at the Crossroads," *Beijing Review,* Vol. 48, No. 38, September 2005. Lorenz, Andreas, "The Chinese Miracle Will End Soon: Spiegel Interview with China's Deputy Minister of the Environment," Patrick Kessler, trans., *Der Spiegel,* No. 10, March 7, 2005.

Lowe, Robert, and Claire Spencer, *Iran, Its Neighbours and the Regional Crises*, Middle East Programme Report, London: Chatham House (The Royal Institute for International Affairs), 2006.

Ma, Jun, *China's Water Crisis [Zhongguo Shui Weiji]*, Norwalk, Conn.: Eastbridge, 2004.

Maloney, Suzanne, "Agents or Obstacles? Parastatal Foundations and Challenges for Iranian Development," in Parvin Alizadeh, ed., *The Economy of Iran: The Dilemmas of an Islamic State*, London: I. B. Tauris, 2001.

Manion, Melanie, *Corruption by Design: Building Clean Government in Mainland China and Hong Kong*, Cambridge, Mass.: Harvard University Press, 2004.

McGregor, Richard, "China Must Cut Farming Population, Says OECD," *Financial Times* (London), November 14, 2005.

————, "China to Drop Rigid Ambitions for Growth," *Financial Times* (London), March 7, 2006.

————, "750,000 a Year Killed by Chinese Pollution," *Financial Times* (London), July 2, 2007.

Mehrara, Mohsen, "Energy-GDP Relationship for Oil-Exporting Countries: Iran, Kuwait and Saudi Arabia," Organization of the Petroleum Exporting Countries, March 2007.

Messkoub, Mahmood, "Social Policy in Iran in the Twentieth Century," *Iranian Studies*, Vol. 39, No. 2, June 2006.

Miller, James, *Review of Water Resources and Desalination Technologies*, Sandia National Laboratories, March 2003.

Movaghar, Afarin Rahimi, Ali Farhoodian, and Reza Rad Goodarz, "A Qualitative Study of Changes in Demand and Supply of Illicit Drugs and the Related Interventions in Bam During the First Year After the Earthquake," Iranian National Center for Addiction Studies, Winter 2005.

National Bureau of Statistics of China, *China Statistical Yearbook*, Beijing: China State Statistical Press, 2005.

————, *China Statistical Yearbook 2006*, Beijing: China State Statistical Press, 2006.

National Oceanic and Atmospheric Administration Satellite and Information Service, "Billion Dollar U.S. Weather Disasters," last updated on July 22, 2008. As of September 12, 2008:
http://www.ncdc.noaa.gov/oa/reports/billionz.html

National Petroleum Council, *Facing the Hard Truths About Energy: A Comprehensive View to 2030 of Global Oil and Natural Gas*, July 18, 2007.

Naughton, Barry, *The Chinese Economy: Transitions and Growth*, Cambridge, Mass.: MIT Press, 2007.

"Nuclear Power Plants Will Generate 6,000MW by 2010," *Iran Daily*, November 7, 2005.

Office of Management and Budget, *Historical Tables, Budget of the United States Government, Fiscal Year 2007*, Washington, D.C.: U.S. Government Printing Office, 2006.

Ogden, Doug, *China's Energy Challenge*, The Energy Foundation, 2004.

Organisation for Economic Co-operation and Development, *Governance in China: Fighting Corruption in China*, Paris, 2005a.

———, *OECD Economic Surveys: China*, Vol. 2005, No. 13, September 2005b.

———, *OECD Economic Surveys: Russian Federation*, Vol. 2006, No. 17, Paris, 2006.

Pei, Minxin, "Corruption Threatens China's Future," Carnegie Policy Brief 55, October 2007.

People's Republic of China National Population and Family Planning Commission, home page, copyright 2001.

Pollack, Kenneth M., "Iran: Three Alternative Futures," *The Middle East Review of International Affairs*, June 2006.

"Pollution Hits China Farmland," BBC News, April 23, 2007.

Population Reference Bureau, "Iran," Web page, undated. As of May 8. 2007: http://www.prb.org/Countries/Iran.aspx

"Power Coal Reserve Falls to 12 Days Amid Rising Prices," *China Daily*, April 23, 2008.

Rahimi, Babak, "Iran: The 2006 Elections and the Making of Authoritarian Democracy," *Nebula*, Vol. 4, No. 1, March 2007.

Rawski, Thomas G., "Recent Developments in China's Labour Economy," November 20, 2003. As of May 30, 2008: http://www.soc.duke.edu/sloan_2004/Papers/Rawski_China_papertext_Nov03.pdf

"Reaching for a Renaissance: A Special Report on China and Its Region," *The Economist*, March 31–April 6, 2007.

Reynolds, James, "Wifeless Future for China's Men," BBC News, February 12, 2007.

Rivlin, Alice M., and J. R. Antos, *Restoring Fiscal Sanity: The Health Spending Challenge*, Brookings Institution Press, 2007.

Russian Ministry of Energy, *Energy Strategy of Russia for up to 2020*, Moscow, 2003.

Samii, Bill, "Iran: Weak Economy Challenges Populist President," Radio Free Europe/Radio Liberty, July 21, 2006.

"SARS Impact Serious but Not Overwhelming: APEC Report," *Xinhua News Agency*, October 17, 2003.

Savadove, Bill, "SARS-Wary Villagers Riot, Attack Officials," *South China Morning Post*, May 6, 2003.

Sawhill, Isabel, and John E. Morton, *Economic Mobility: Is the American Dream Alive and Well?* Washington, D.C.: The PEW Charitable Funds, 2007.

Sepehri, Vahid, "Iran: Officials Debate Rate Cuts to Curb Inflation," Radio Free Europe/Radio Liberty, May 23, 2007.

Shafie-Pour, Majid, and Mojtaba Ardestani, "Environmental Damage Costs in Iran by the Energy Sector," *Energy Policy*, April 2007.

Shalizi, Zmarak, "Addressing China's Growing Water Shortages and Associated Social and Environmental Consequences," World Bank Policy Research Working Paper 3895, April 2006.

Shi, Jiangtao, "Pollution Makes Yangtze 'Cancerous,'" *South China Morning Post*, May 31, 2006.

Shrestha, Laura B., *The Changing Demographic Profile of the United States*, Congressional Research Service, RL32701, updated June 7, 2006.

Smil, Vaclav, *China's Past, China's Future: Energy, Food, Environment*, New York: Routledge Curzon, 2004.

Social Security Administration, *The 2002 Annual Report of the Board of Trustees of the Federal Old-Age and Survivors Insurance and Disability Insurance Trust Funds*, March 26, 2002.

————, *The 2007 Annual Report of the Board of Trustees of the Federal Old-Age and Survivors Insurance and Disability Insurance Trust Funds*, May 1, 2007.

Stern, Roger, "The Iranian Petroleum Crisis and United States National Security," *PNAS*, Vol. 104, No. 1, January 2, 2007.

"Supply Crunch Hits Coal Cradle," *China Daily*, July 9, 2008.

"Third of China 'Hit by Acid Rain,'" BBC News, August 27, 2006.

Thurow, Lester, "A Chinese Century? Maybe It's the Next One," *New York Times*, August 19, 2007.

Torbat, Akbar E., "The Brain Drain from Iran to the United States," *Middle East Journal*, Vol. 56, No. 2, Spring 2002.

Tu, Jianjun, "China's Botched Coal Statistics," *China Brief*, Vol. 6, No. 21, October 25, 2006.

Tucker, Patrick, "Urbanization Models Examined," *The Futurist*, Vol. 39, No. 6, Nov/Dec 2005.

Unger, Jonathan, "China's Conservative Middle Class," *Far Eastern Economic Review*, April 2006.

United Nations Population Division, "World Population Prospects: The 2004 Revision Population Database," 2004.

————, "World Population Prospects: The 2006 Revision Population Database," last updated September 20, 2007. As of September 1, 2007: http://esa.un.org/unpp

U.S. Census Bureau, Population Division, *International Data Base*, last updated June 16, 2008. As of September 1, 2007: http://www.census.gov/ipc/www/idb

U.S Department of Commerce, Census Data, 2007.

U.S. Government Accountability Office, *Issues Related to Potential Reductions in Venezuelan Oil Production*, GAO-06-68, June 2006.

————, *Uncertainty About Future Oil Supply Makes It Important to Develop a Strategy for Addressing a Peak and Decline in Oil Production*, GAO-07-283, February 2007.

"U.S.-Japan Defense Alliance Strengthens," *Honolulu Advisor*, June 3, 2007.

Valadkhani, Abbas, "What Determines Private Investment in Iran?" *International Journal of Social Economics*, Vol. 31, No. 5/6, 2004.

van der Veer, Jeroen, "Shell Global Scenarios to 2025," The Hague: Royal Dutch/Shell Group, June 2005.

van Groenendaal, Willem J. H., and Mohammad Mazraati, "A Critical Review of Iran's Buyback Contracts," *Energy Policy*, Vol. 34, 2006.

Verma, Shiv Kumar, "Energy Geopolitics and Iran-Pakistan-India Gas Pipeline," *Energy Policy*, Vol. 35, 2007.

Vidal, John, "Dust, Waste, and Dirty Water: The Deadly Price of China's Miracle," *The Guardian* (London), July 18, 2007.

Walsh, Bryan, "The Impact of Asia's Giants," *Time Magazine*, March 26, 2006.

Wang, Tao, and Wei Wu, "Sandy Desertification in Northern China," in Kristen A. Day, *China's Environment and the Challenge of Sustainable Development*, Armonk, N.Y.: M.E. Sharpe, 2005.

Wang, Yan, Dianqing Xu, Zhi Wang, and Fan Zhai, "Implicit Pension Debt, Transition Cost, Options, and the Impact of China's Pension Reform: A Computable General Equilibrium Analysis," The World Bank, Policy Research Working Paper, February 2001.

Watts, Jonathan, "Punch Line: SARS Sparks Chinese Riots," *The Guardian* (London), May 6, 2003.

Wedeman, Andrew, "Budgets, Extra-Budgets, and Small Treasuries: Illegal Money and Local Autonomy in China," *Journal of Contemporary China*, Vol. 9, No. 25, 2000.

Williams, Frances, "China Exports Overtake Japan," *Financial Times* (London), April 14, 2005.

Williams, Stuart, "Ahmadinejad Faces New Setback in Key Poll Battle," *Middle East Times*, December 19, 2006.

Wolf, Charles Jr., K. C. Yeh, Benjamin Zycher, Nicholas Eberstadt, and Sungho Lee, *Fault Lines in China's Economic Terrain*, unpublished RAND Corporation research.

Wonacott, Peter, "Ailing Patient: In Rural China, Health Care Grows Expensive, Elusive," *Wall Street Journal*, May 19, 2003.

Woodrow Wilson Center for International Scholars, "Energy in China Fact Sheet," 2005. As of May 30, 2008:
http://www.wilsoncenter.org/topics/docs/energy_factsheet.pdf

Woodruff, Yulia, "Russian Oil Industry Between State and Market," in *Fundamentals of the Global Oil and Gas Industry, 2006,* Petroleum Economist, October 2006.

World Bank, *World Development Reports, Country Tables: China*, undated. As of June 3, 2008:
http://hdr.undp.org/

————, *Clear Waters, Blue Skies: China's Environment in the New Century*, Washington, D.C., 1997.

————, *Russian Economic Report No. 13*, Washington, D.C., 2006.

————, *Cost of Pollution in China: Economic Estimates of Physical Damages*, Washington, D.C., February 2007.

Yardley, Jim, and David Barboza, "Help Wanted: China Finds Itself with a Labor Shortage," *New York Times*, April 3, 2005.

Zeng, Douglas Zhihua, "China's Employment Challenges and Strategies After the WTO Accession," World Bank Policy Research Working Paper 3522, February 2005.